CONTENTS

CONTENTS

Creative Techniques
in Product and Engineering Design
- A Practical Workbook

D J Walker
Lecturer, Design Discipline, Faculty of Technology,
The Open University

B K J Dagger
Industrial Consultant, Design and Manufacture
(Formerly, Project Manager - Training Development, EITB)

R Roy
Senior Lecturer, Design Discipline, Faculty of Technology,
The Open University

WOODHEAD PU

Published by
Woodhead Publishing Ltd, Abington Hall, Abington,
Cambridge CB1 6AH, England

First published 1991

British Library Cataloguing in Publication Data
A CIP catalogue record for this book is available from the British Library.

ISBN 1 85573 025 1

Foreword

My Institution has long realised that Engineering Design as practised is, apart from much hard work, dependent upon strict disciplines. The best ideas and designs are rarely the result of gazing into space, but rather are the result of the systematic use of thinking skills. This book introduces, helps and guides the student through a variety of approaches to solving design problems. The Institution of Engineering Designers recommends the book to all those who would wish to make a career in this noble and exciting profession.

Dennis V Shepherd
Past President
The Institution of Engineering Designers

Preface

If you have the task of teaching design to students in secondary, further or higher education, then providing each student with a copy of the book and your tutorial support based upon the teaching points made in the Study Text will provide you with an extensive teaching and training resource. We recommend studying very carefully each section before presenting a lesson. You should be the best judge of how to tutor your own particular students. However, we intended that the 'Activities' and 'Work Tasks' be used for student activities. The Study Text should provide you with the material necessary for any formal teaching prior to, during or after, the student activities.

INTRODUCTION

For many people there is something mysterious about creativity - about the ability of artists, engineers and designers to create something solid and convincing out of thin air. It seems like a conjuring trick, available only to those in the 'magic circle'. For them, the indefinite space of possibilities and the blank page - far from stimulating - is an intimidation which brings paralysis.

This is in dramatic contrast to designers, engineers and scientists - the professionals who revel in the challenge of the blank page and what they call *blue skies* projects. Such projects seem to be an invitation to do what they like. The implication, of course, is that creative abilities work best when unconstrained.

However, from your own experience you will know that unfettered creative opportunities are rare in the world. More than that, you may know that even the most creative people use a framework and techniques to help them in their work. There are a number of well tried techniques which can act as a stimulus to exploring problems. Some techniques are buried in implicit designer folklore but some are explicit and relatively easy to follow and use.

> *So the attitudes embodied in this book are:*

- that creativity is not particularly mysterious
- that creativity does respond to a framework and a system
- that there are well established techniques which foster creativity

We have chosen only a few *creative techniques* (eight to be precise) to cover here. This means that we are able to deal with them in some depth and give the room for them to be rehearsed. Think of them as utilitarian tools which are not infallible. We recommend that the techniques be used conscientiously as if by a worker, and not self-consciously as if by an artiste or conjurer. Using them in the right way, and in the right frame of mind, can open up wider and unexpected possibilities.

The book aims to help improve design abilities. You are shown how to apply a variety of creative techniques to design problems. For design lecturers wishing to use the book as a teaching resource, detailed reference to its use as self-study learning materials is made in the User Guide.

The structure of the book is as follows:

USER GUIDE

This gives detailed advice about how the study materials can be used to best effect.

STUDY TEXT

Section 1 - The Nature of Design Thinking outlines an overall view of the process of design; and examines the *styles of thinking* necessary to solve different types of design problem.

Section 2 - Creative Design Techniques examines the problem solving process of design and shows how eight different *creative techniques* can be applied to design problems.

Section 3 - Creative Approaches to Design Problems helps in the selection of the most appropriate techniques for solving design problems.

USER GUIDE

WHO CAN BENEFIT FROM THE STUDY MATERIALS ?

The study materials are directed at all those who have to apply or teach design skills. There is a fairly strong emphasis on design as it impinges on products which have to be 'engineered' and require 'technological processes' for their manufacture. But most of the techniques are useful in design tasks in other areas of design, such as furniture, architecture, building design, the process and service industries - examples from all these areas are used.

DESIGN STUDENTS If you are studying design at a college of further or higher education and wish to develop your design skills prior to a career as a designer, then working through the book and obtaining support from your college tutor should give you some good design practice and develop your skills.

DESIGNERS If you are a practising designer and wish to enhance your skills in ways which will benefit you and your company on future design projects, then working through the book and applying the techniques to job related tasks should stimulate your design ideas.

DESIGN MANAGERS If you are responsible for young designers and you wish to stimulate their design ideas, then providing each person with a copy of the book and with your tutorial support (or that of a senior designer) should provide an effective answer.

WHAT DOES THE BOOK CONTAIN ?

INTRODUCTION
USER GUIDE
STUDY TEXT which includes:
Study Guide
Objectives
Examples
Case Studies
Activities
Discussions of Activities
Work Tasks

HOW DO YOU START ?

SPEND AT LEAST 30 MINUTES THUMBING THROUGH THE BOOK - TAKING PARTICULAR NOTICE OF THE WORK TASKS - BEFORE READING ANY FURTHER

DESIGN STUDENTS If possible, show this book to your design lecturer at college and ask him/her to act as your tutor whilst studying the book. The lecturer will probably want to know what is expected, so you will need to agree about what he/she is prepared to do - it makes sense to try to relate this to his/her existing role as your lecturer and the design syllabus being followed. Ideally, you should get him/her to agree to comment upon the 'Activities' you have undertaken and to set 'Work Tasks' which relate to your studies.

DESIGNERS If possible, discuss with your manager the issue of who is to act as your tutor. Having established this, and in the presence of your tutor, discuss those aspects of your job where you need to develop your skills. This may involve enhancing your design skills in ways which will benefit you and your company on future design projects. Ideally, the outcome of these discussions should result in agreements on work related tasks - known in the Study Text as 'Work Tasks'.

HOW DO YOU PROCEED ?

For most people, it is best to start at the beginning of the Study Text and work progressively through the materials step by step. However, for people such as practising designers, you may choose to apply only the techniques which seem appropriate to your design problems.

HOW MUCH TIME WILL BE INVOLVED ?

This will vary greatly, depending upon the amount of actual design project work you choose to undertake - resulting from the Work Tasks. Otherwise, it will depend upon who you are, for instance:

DESIGN STUDENTS If you have a copy of the book and you are using it as self-study materials, then the time necessary to study the Study Text and undertake the associated Activities will depend upon your present knowledge of the process of design. But, we anticipate that it will take from 30 to 40 hours plus the time to undertake the Work Tasks.

DESIGNERS The time necessary to study the Study Text and undertake the associated Activities will depend upon your present knowledge of the techniques presented. But, we anticipate that it will take from 20 to 30 hours plus the time to undertake the Work Tasks.

DESIGN LECTURERS / TEACHERS The time necessary to study the Study Text and prepare your lessons will depend upon your present knowledge of the techniques presented, and the syllabus you are following. However, we anticipate that to study the Study Text will take you from 15 to 25 hours. Your students' study time involved in undertaking each Activity is recommended in each case - but you will need to make your own judgement on the basis of the abilities and previous knowledge of your students. In addition, your judgement will be particularly necessary when allowing sufficient time to undertake Work Tasks.

WHAT ARE 'ACTIVITIES' ?

The Activities:

- test the understanding of the principles, concepts and ideas presented
- allow the creative techniques presented to be practised

It is recommended that each Activity be undertaken immediately after studying the relevant text, or, in the case of lecturers, having taught the relevant subject matter. In most cases there is more than one solution to the Activity concerned. Following each Activity there is a 'Discussion of Activity' which helps you assess your understanding. For lecturers, the suggested solutions to Activities should form a useful focus for tutorial sessions.

WHAT ARE 'WORK TASKS' ?

The Work Tasks:

- give practice in the use of the principles, concepts and ideas presented
- allow the creative techniques presented to be practised - thereby developing design skills by tackling design problems in new ways.

It is recommended that each Work Task be undertaken immediately after undertaking the related Activities, or, in the case of lecturers, having taught the relevant subject matter. The Work Tasks should form useful project based activities.

WHAT IS A WORK FILE ?

As a result of undertaking the Activities and Work Tasks various pieces of work will of course be generated. The *Work File* should be quite simply wherever you feel is the most appropriate place to keep this work. It may be a ring-binder, folder or sketch pad specifically set aside for the purpose.

The contents of the Work File may include written notes, calculations, tables, graphs and various design communications, including sketches, design concept drawings and detail drawings. Ideally, the Work File will be a part of, or develop into, a portfolio of work which properly reflects the design abilities being exercised.

The Tutor may wish to look at the Work File in order to make detailed comments about the Activities and Work Tasks undertaken. These comments should seek to increase students' understanding of the principles and concepts applied, and stimulate further investigation into the topics involved.

WHAT ARE THE AIMS OF THE WORK FILE ?

To show that each of the Objectives have been met by:

- demonstrating the use of the techniques and practices presented

- recording the responses to the Activities in the Study Text

- recording and reporting upon the Work Tasks undertaken

- providing a source of reference for future design studies.

INDIVIDUAL OR GROUP ACTIVITIES ?

All the creative techniques presented can be used by individuals - working by themselves - to stimulate their thinking and thereby possibly give them new ideas. But, if creativity (as we shall later discuss) is associated with the ability to merge previously unconnected pieces of knowledge and experience, then the potential of a group of people to be creative must be larger than the individual's potential. A group of people - especially from different backgrounds and disciplines - provide a greater pool of knowledge and experience to call upon. Some of the techniques presented are systematic approaches which set out to increase the possibility of bringing together completely different elements of knowledge in seeking new solutions to design problems.

Whilst the book contains self-study activities which can be used by individuals working on their own, it is recommended, where appropriate, that groups be used. For creative stimulation to take place in a group, it needs the effective exchange of each member's knowledge and experience. This requires the use of an informal and relaxed style of communication. Experience has shown that the optimum size of such groups should be between five and seven participants. In groups larger than this, the individual members may become inhibited and consequently their capabilities will not be properly employed. With smaller groups, the problem becomes one of too little mutual stimulation operating on a smaller base of knowledge.

SECTION 1

The Nature of Design Thinking

What is a design?

That is a question that can spark-off hours of debate!

People's answers can range from the methodical:

'The optimum solution to the sum of the true needs of a particular set of circumstances'.

(Edward Matchett)

to the mystical:

'The performing of a very complicated act of faith'.

(J. Christopher Jones)

This first section of the Study Text does not attempt to join this debate! Instead, it concentrates firstly on outlining an overall view of the process of design that I hope will seem familiar to you. Then it goes on to consider some important aspects of the natural tendencies that we all seem to have in our 'styles of thinking'. Understanding these styles - and your own preferred style in particular - lays a foundation for the later, main section of the Study Text on 'creative' design techniques.

OBJECTIVES

After completing the study materials on this section you will be able to:

1. Describe the general nature of the design process and outline its main phases.

2. Explain how different 'styles' of thinking are appropriate to solving different types of design problem.

3. Identify your own preference for using a 'convergent' or 'divergent' thinking style.

Design Process

I imagine that in the kind of design work with which you are familiar, the early part of the design process is characterised by high uncertainty and a sometimes vague list of requirements; whilst the end point of the process is a technical specification for a single component or assembly where the levels of accuracy and certainty are high. We can say, then, that the task which is accomplished by the design process is to convert 'loose' problems into 'tight' solutions. This is clearly a *converging* process.

In the early stages of the process, your role may require you to bring the main requirements into focus. This often means challenging the assumptions of the design 'brief', or generating wide-ranging hypothetical design ideas - which are perhaps not so much realistic solutions but more a testing out of the parameters of the problem. So within the design process, and particularly in the early stages, there are times when a *diverging* process is necessary.

> *In the process of converging to a single design or single family of designs there are probably two main questions that you face:*

> 'Have I overlooked any alternatives?' (Diverging)

> 'Is this a workable version of the solution?' (Converging)

> *I propose to examine the flow of activity from an indefinitely large number of design possibilities to a single final object by considering the process as a movement through three phases:*

- concepts

- embodiment

- details

No doubt you have your own words to describe these three design phases as they occur in your own situation.

THREE PHASES OF DESIGN CONVERGENCE

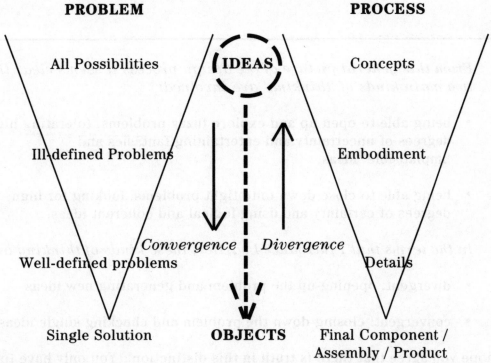

PROBLEM   **PROCESS**

All Possibilities   (IDEAS)   Concepts

Ill-defined Problems   Embodiment

Convergence | *Divergence*

Well-defined problems   Details

Single Solution   **OBJECTS**   Final Component /
                                          Assembly / Product

Concepts

My interpretation of the design process is that the first phase deals with
very general, sometimes vague, and sometimes wild ideas. The culmination
of this phase is an outline of a target area to work within. However, the
parameters are still fluid, the design ideas are unstable and the problems
are fuzzy. Many ideas are still under consideration.

Embodiment

The next phase engages with principles of arrangement geometry, and
major elements. Your task is to manipulate these properties until a
preferred form emerges. One idea or set of ideas dominates from this
point. (The Ford Motor Company calls this point of the design process
'Go with one').

Details

The final phase is one of definition within clearly established boundaries.
The details of design, the attributes of components and manufacturing
processes are brought into focus - giving an appropriate single solution.

I imagine that these three phases correspond in a general way to your own
overall *process of convergence* in coming to design solutions.

Design Thinking

From this general picture of the design process it seems clear that two main kinds of 'thinking' are involved:

- being able to open up and explore fuzzy problems, tolerating high degrees of uncertainty and entertaining fantasies and 'implausible' ideas.

- being able to close down onto tight problems, looking for high degrees of certainty and using logical and coherent ideas.

In the terms that I introduced earlier, these kinds of thinking are:

- divergent: opening-up the problem and generating new ideas

- convergent: closing-down the problem and checking single ideas.

I hope you agree that there is truth in this distinction! You only have to think of the tensions that exist between strategic planning and detailed tactics - between the beginning and the end of the design process. Divergent thinking is more appropriate in the early stages of design, whereas convergent thinking is more useful in the later stages.

Although this difference in the appropriate kinds of thinking is a key aspect of the nature of the design process, it is also a reflection of basic differences in preferred styles of thinking between individuals. I assume that you, like most people, are able to operate in both modes - divergent and convergent - but that you probably prefer, or feel happier with, one mode rather than the other.

Due to this natural preference, and because designing requires *both* thinking styles, I hope to encourage you to take on ways of thinking which are not always natural to you - to make you more 'convergent' if you are a 'diverger', and more 'divergent' if you are a 'converger'. Techniques for developing and improving these thinking styles are discussed in detail in Section 2 of the Study Text.

The looser, more informal techniques that tend to be most appropriate at the early phases of the design process are a stimuli to *divergent* thinking. Tighter, more formal methods are often most useful in the middle and later phases of the design process. Most of them are stimuli to *convergent* thinking.

Your own thinking style

On the following pages you will find ten problems which are intended to make you more sensitive to your own style of thinking. Following this, we go on to examine which styles of thinking are appropriate to different design problems.

Look on these Activities as exercises in *self-discovery*.

As you work through them, think about the kind of problem you feel at ease with and the kind of problem which irritates you. Think about the medium you prefer to work in, e.g. words, diagrams or drawings.

The problems used in these Activities have been derived from various sources. They cover a typical range of skills from visual perception, to understanding construction, reading ambiguities, deducing underlying rules and so on.

Do not be disappointed if you do not produce a correct answer to each problem. For some of them, there are no 'correct answers', only a range of alternatives.

If you do well in half of the exercises, then you are demonstrating thinking skills which are valuable for solving design problems. I hope to show you that trying these problems and puzzles is not a trivial exercise - they are to do with real designing. The Activities should help you discover something about your own thinking skills. And hence, which of the styles of thinking are most appropriate to solving different types of design problem.

You should attempt each of these ten activities by entering your responses in your Work File. The Discussion of Activities 1 to 10 which follows will assume that they have been attempted.

So at this point I would like you to review your own preferred methods of working, and to come to an understanding of your own preferred 'style' of thinking.

Which of the following statements do you think describes your ways of working and thinking?

'I am methodical' or 'I think in fits and starts'.

'I like to be orderly' or 'I start at the most important issue'.

'I like to deal with clear objectives' or 'I like to make bold sketches of whole solutions'.

Perhaps you don't see yourself so easily categorised! So let's investigate your style of thinking a bit more thoroughly. Try the following activities. They are not tests of 'intelligence'; don't worry about finding exactly correct answers because very often there is a range of answers.

I would like you to spend 3-5 minutes on each Activity and enter your responses in your Work File.

ACTIVITY 1

Here are two views of a real three dimensional object.
What shape is it? Draw the complete shape.

| Side View | Front View |

ACTIVITY 2

How many ways can you interpret or 'see' this drawing?
Use a phrase for each interpretation (e.g. box with steps cut out of it) and make a drawing (with labels if necessary) of each so as to make your interpretation visually clear.

ACTIVITY 3

Two blocks of wood are dovetailed together as shown. The faces farthest away are identical to those shown.
How were these pieces put together? Draw a sketch to explain.

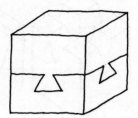

ACTIVITY 4

It is possible to join all these dots together in four connected straight lines, without taking your pen from the paper or going back over a line. How?

ACTIVITY 5

Three jugs, A, B and C are to be used to measure out a given volume of liquid. The table below sets out a number of problems with jugs of differing sizes, and the required volume to be measured out in each case. For example, the solution to problem 1 can be achieved by filling jugs A and B, pouring out enough from jug A to fill jug C, and then adding the contents of jug B to jug A; the answer is thus A - C + B. Try to solve the remaining problems as quickly as you can, aiming to achieve the required volume with as little pouring back-and-forth between jugs as possible.

Problem No.	Volume of jugs			Volume required	Answer
	A	B	C		
1	10	7	8	9	A - C + B
2	25	20	11	16	_____
3	14	3	2	13	_____
4	18	10	7	15	_____
5	11	8	6	9	_____
6	23	18	9	14	_____

ACTIVITY 6

What is the next shape in the sequence below?

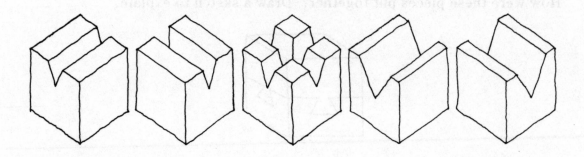

ACTIVITY 7

What are the next two shapes in the sequence below?
Explain why.

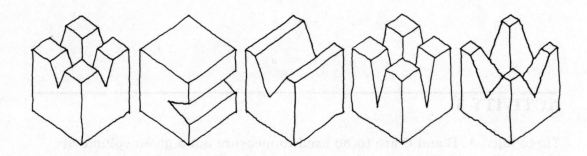

ACTIVITY 8

Draw the next square in this series.

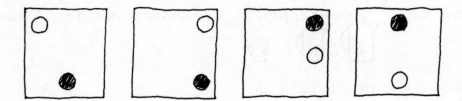

ACTIVITY 9

List as many things that you can think of that this drawing might represent.

ACTIVITY 10

With which of the following statements do you agree?
Make a note of your four preferred statements.

Designing is .....

a) *going for a walk in a maze*

b) *applying scientific and mathematical principles to the solution of practical problems*

c) *resolving conflicting requirements*

d) *finding an unexpected answer*

e) *assembling known components in a new order*

f) *matching the needs of people to the resources of technology*

g) *decision-making in the face of uncertainty*

h) *creating something new*

Note Spend 3-5 minutes on each Activity and enter your responses in your Work File.

DISCUSSIONS OF ACTIVITIES 1 TO 10

DISCUSSION OF ACTIVITY 1
Projection

Let us compare your responses with the answers given below:

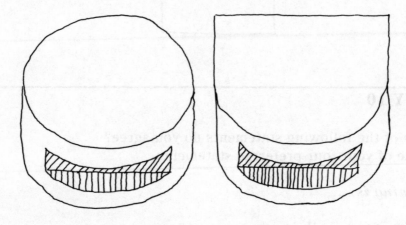

Two answers to Activity 1

I expect you got one of the answers to Activity 1, but did you get both? Solving such problems involves the ability to see objects represented in two dimensions, in three dimensions and vice versa. From this comes the ability to draw and sketch two and three-dimensional objects.

Such skills are as important in design as the ability to calculate the stress in a component or select a suitable material for a product. They are particularly important at the early conceptual stage of design.

DISCUSSION OF ACTIVITY 2
Perception

Look now at the answers to Activity 2 given below. Here, I have given three interpretations of the drawing. You may have given other interpretations since these are not the only ones. These illusions work by ambiguity and visual reversals.

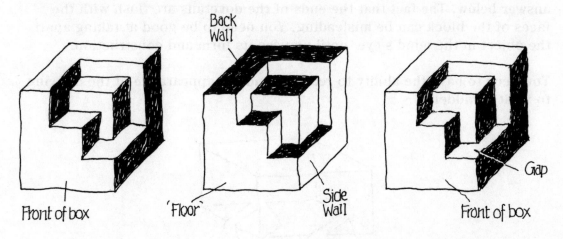

a) Box with steps cut out of it

b) Corner seat in a room

c) Box with steps and 'L' shaped solid suspended above the steps

Three answers to Activity 2
(There are further interpretations not shown here)

Normally, I see the drawing as a cube with steps cut out of it - object A.

To visualise object B - the corner seat in a room - I have to defocus my eyes, or blink and look again. The alternative shape may then suddenly appear.

The third interpretation, object C, is just a variation on the first. The 'L' shaped piece is floating in mid-air instead of being attached to the lower step.

It's rather like the flash of insight one sometimes gets when the solution to a design problem unexpectedly occurs to you. It clicks into place.

After you've seen it, you wonder why you didn't think of it before!

DISCUSSION OF ACTIVITY 3
Construction

Activities 1 and 2 were exercises in visual perception.

So is Activity 3 - the block of wood dovetailed in a seemingly impossible way. Yet you can visualise that the dovetails are diagonal - as shown in the answer below. The fact that the ends of the dovetails are flush with the faces of the block can be misleading. You need to be good at taking apart the object in the mind's eye until you see its form and construction.

You need to have the ability to read beyond the appearance of the drawing to what is hidden.

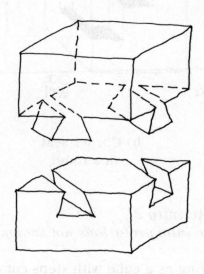

DISCUSSION POINTS FOR ACTIVITIES 1 TO 3
Visualisation Skills

Visualisation skills extend beyond being able to 'see' a shape or form, into being able to envisage how forces are transmitted within an object being designed. For example, where stress concentrations occur, or how components interact, and even to 'see' the flow of fluids, electromagnetic fields and so on inside a piece of equipment.

One designer I know says, "Before putting anything on paper, I lie in the bath and imagine the component I am designing being produced on a lathe or other appropriate manufacturing equipment". This imaginative visual ability is very difficult to teach and these study materials do not attempt to do so. What the materials do attempt to do is to stimulate your imaginative skills.

In Section 2 of the Study Text there are specific techniques to help stimulate your imaginative and inventive abilities.

For instance the analogies technique involves you in imagining yourself inside the object being designed, experiencing the forces, flows and motions involved.

Or, the new combinations method gets you to deliberately combine two or more previously unrelated objects in order to produce a new design concept.

Think for example of a combination of a ball shaped wheel with a wheelbarrow to produce an improved design - the ball barrow!

DISCUSSION OF ACTIVITY 4
Boundaries

Let's look next at the solution to Activity 4 given below. You may have seen this nine dots problem before. It's in almost every book on creativity ever published and so you may have remembered the answer. The reason it's so widely used is that it illustrates a very important point about creative design - the need to break out of artificial boundaries or constraints.

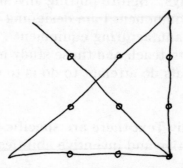

The solution is obtained if you do not assume that the four lines must be within the boundary of the nine dots.

In this case, you can only draw a continuous line linking all nine dots, without lifting your pencil from the paper, if you realise that you do not have to stay within the boundary formed by the eight edge dots. Often we prevent ourselves solving a design problem by assuming that a constraint exists when it is only in our minds. This problem is sometimes referred to as the 'Eskimo problem' - apparently Eskimo children find it very easy - perhaps because they do not experience boundaries to their physical environment as we do.

Later on in Section 2 of the Study Text, the Boundary Shifting technique shows how relaxing unnecessary constraints can often enable better solutions to design problems to be made - while staying within the necessary constraints of the Designer's 'brief'.

DISCUSSION OF ACTIVITY 5
Logic

Activity 5, the three jugs problem, may seem more like a logic puzzle than anything to do directly with creative design. But what was your answer to the last sub-problem number 6?

Problem No.	Answer
1	A - C + B
2	A - B + C
3	A - B + C
4	A - B + C
5	A - B + C
6	A - C

Most people give the same answer as to sub-problems 2 to 5, that is, **A - B + C**. This is not wrong but it is not the best or most elegant solution. If you put **A - C**, then you are to be congratulated because your thinking was not stuck in a rut from doing the previous four Activities.

DISCUSSION POINTS FOR ACTIVITIES 4 & 5
Traditional Solutions

Being in a rut is often the result of tackling design problems under severe time constraints. We often tend to stick to familiar ideas and known solutions when we are under pressure.

Time pressure and known solutions are of course very common in design and are not necessarily a bad thing. But the habits and ways of thought conditioning by tradition and experience can be a major barrier to innovation. As a result, traditional solutions seem inevitable.

For example it took nearly twenty five years for car designers at the turn of the century to break away from the horse drawn cart type of chassis. Early car designers forgot that a horse pulled a cart, while a motor vehicle was pushed by its wheels, producing substantial twisting forces.

It was another forty years before someone realised that the engine could be placed transversely in the body - with the gearbox out of line with the engine's crankshaft instead of always in-line.

DISCUSSION OF ACTIVITY 6
Combination

Now look at the solution to Activity 6 given below. Getting the answer requires 'step by step' *convergent* thinking - the kind of thinking required when analysing an engineering problem with only one correct answer. If you did not get the answer given, look again at how the first two shapes are combined to make the third. The solution is obtained by combining the two shapes in the next sequence of three. This seems straightforward but it has an implication for Activity 7.

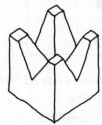

Answer requires logical, convergent thinking

DISCUSSION OF ACTIVITY 7
Deduction

Did you manage to solve Activity 7? It looks like the same type of problem as before. But to come up with the answer given you have to break out of the linear way of thinking you used in Activity 6. Instead of looking at the sequence 'step by step', you have to view it as a whole. Activity 6 in a sense produces the wrong frame of mind (or psychological set) for Activity 7.

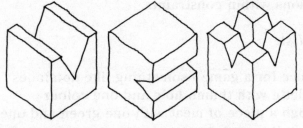

Answer requires lateral, divergent thinking
(The sequence is meant to be chess pieces)

Once you realise that the shapes are meant to be chess pieces, with a rook (or castle) on the left, and the king on the right, it is easy to see a solution. Shown above are the next three shapes - representing a bishop, a knight and a rook. This is not meant to be a trick question, and you may have come up with a different but quite valid solution. It is meant to show that the kind of *lateral thinking* stimulated by viewing something as a whole from outside, in this case pieces for a game, is an important skill in creative problem solving.

DISCUSSION OF ACTIVITY 8
Convergence

Activity 8 goes back to the logical and *convergent thinking* of Activity 6. You have to work out the sequence with which the dots move around the square - as if in a 3 X 3 matrix. The black dot moves anti-clockwise and the white dot moves clockwise.

Answer requires logical, convergent thinking
(Black dot moves anticlockwise $\frac{1}{2}$, 1, $\frac{1}{2}$, 1 side, white dot
moves clockwise 1, $\frac{1}{2}$, 1, $\frac{1}{2}$ side)

DISCUSSION OF ACTIVITY 9
Divergence

In contrast, Activity 9 is typical of the type of task set to assess *divergent thinking*. No doubt your responses are different to those given below because, of course, there are no right answers. What is more important is the number and variety of responses. If you produce lots of responses, including some 'way-out' ideas, you're probably a naturally divergent thinker. If you produced a few of the more obvious ideas, your thinking style may be more convergent - better at analysing problems and working out optimum solutions within constraints.

Here is my list:

- Playing piece for a game - something like dominoes
- Artist's palette with thumb-hole and one colour
- Slice through a piece of meat with one green and one black olive embedded in it
- Piece of Swiss cheese with a worm in one hole
- Front view of a camera
- Cuff-link with black and white jewels
- Door with a viewing panel in it
- Swimming pool with two circular rafts
- Switch with on and off buttons
- Table with glasses of wine - one red, one white
- etc, etc

DISCUSSION OF ACTIVITY 10
Metaphor

Finally, as a summary of your own ideas of design, compare your responses to Activity 10 with my comments given below. You probably didn't fall neatly into either of the two categories given because, at different times, solving design problems requires both convergent and divergent thinking. At times it may be like walking in a maze and at others a matter of detailed mathematical analysis. The point of attempting these activities is to see which style of design thinking you find easiest and which you find more difficult, so that you can be more sensitive to developing and exploiting your skills.

Designing is ......

Answers that suggest a preferred flexible / divergent thinking style:	Answers that suggest a preferred focused / convergent thinking style:
a) *going for a walk in a maze* d) *finding an unexpected answer* f) *matching the needs of people to the resources of technology* h) *creating something new*	b) *applying scientific and mathematical principles to the solution of practical problems* c) *resolving conflicting requirements* e) *assembling known components in a new order* g) *decision-making in the face of uncertainty*

Now read the Discussion of Activities 1 to 10 and consider whether you have a preferred 'style of thinking'.

DISCUSSION OF 'YOUR OWN THINKING STYLE'

In general, these activities explore two modes of thought (see table below 'Styles of Thinking') - but it is rare that anyone will belong exclusively to one mode or the other. In engineering design there is an oscillation in the design process from one mode to another - pulling things apart and putting them together, or, more grandly, analytic and synthetic. Because the design process as a whole is converging, there is a natural shift from open flexible thinking at the beginning to narrower focused thinking at the end. So neither mode of thought is better than the other, but each has its own appropriate place in the design process. At certain points in the process it is foolish to open up more fundamental possibilities - at other points it is equally foolish to work in too narrow or too precise a manner.

I hope that you can see the strengths and weaknesses of both styles of thinking - and are encouraged to think, when it is appropriate, in ways that may not, at first, seem natural or preferable to you.

The following table illustrates some of the natural dichotomies which exist in different 'styles of thinking':

Convergent	*Divergent*
Seeks to close down on problem	Seeks to open up more solutions
Works connectedly	Works disconnectedly
Problem orientated	Solution orientated
Rational / deliberate	Intuitive / arbitrary

Focused	*Flexible*
Tight	Loose
Precise	Diagrammatic / sketchy
Deals with single objectives	Multiple objectives
Does not like ambivalence	Accepts ambivalence

Linear	*Lateral*
Seeks correct answers	Seeks appropriate answers
Rigid perceptions	Alternative perceptions
Sticks to rules	Looks for changes of rules
Fixed territory	Boundary shifting

Serialist	*Holist*
Works in sequence	Works in any order
Step by step	Looks for whole picture
Sticks to one viewpoint	Moves through different viewpoints
Moves forward gradually	Steps back: jumps in: steps back

Types of design problem

Let us extend the notion of different thinking styles by looking at different categories of design problem. The following four problems demonstrate quite different kinds of design tasks, moving from a narrow well defined problem to looser more fuzzy problems. Of course, very many problems in design and engineering are like related problems in science and mathematics. They can be solved by using rational analytical thought, by applying known principles to known facts.

ACTIVITY 11
Rotating shaft problem

Look at the shaft design problem illustrated below:

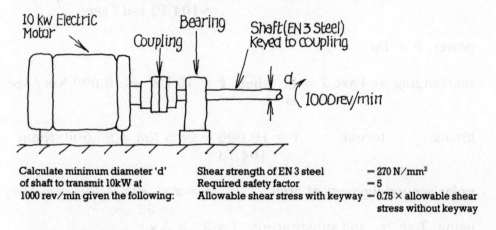

Calculate minimum diameter 'd' of shaft to transmit 10kW at 1000 rev/min given the following:		
Shear strength of EN 3 steel	= 270 N/mm²	
Required safety factor	= 5	
Allowable shear stress with keyway	= 0.75 × allowable shear stress without keyway	

Shaft design problem

(Adapted from Hawkes, B. and Abinett, R. *The enginering design process*, Pitman: London 1984)

Have a go at solving this problem using the formulae provided below and by showing your calculation in your Work File. The reason for introducing this example is to give you a typical engineering problem that involves using mathematical formulae.

Formulae associated with torsion loading in shafts:

$P = T\omega$ where P — power (watts)

T — torque (Nm)

ω — angular velocity (rad./sec)

$J = \dfrac{\pi d^4}{32}$ J — polar second moment of area

d — shaft diameter (mm)

$\dfrac{T}{J} = \dfrac{2\tau}{d}$ τ — allowable shear stress (N/mm²)

DISCUSSION OF ACTIVITY 11
Rotating shaft problem

Your calculation should look something like this:

allowable stress (without keyway)   = $\dfrac{\text{shear strength}}{\text{safety factor}}$

$= \dfrac{270}{5} = 54 \text{ N / mm}^2$

allowable stress (with keyway), $\tau = 0.75 \times 54 = 40.5 \text{ N / mm}^2$

angular velocity, $\omega = \dfrac{1{,}000 \times 2\pi}{60}$

$= 104.73 \text{ rad / sec}$

power, $P = T\omega$

rearranging we have $T = \dfrac{P}{\omega}$, where $P = 10 \text{ kW} = 10{,}000 \text{ Nm / sec}$

giving,   torque $T = \dfrac{10{,}000}{104.73} = 95.5 \text{ Nm} = 95{,}500 \text{ Nmm}$

polar second moment of area, $J = \pi \times \dfrac{d^4}{32}$

using, $\dfrac{T}{J} = \dfrac{2\tau}{d}$ and substituting, $\dfrac{T \times 32}{\pi d^4} = \dfrac{2 \times \tau}{d}$

we have, $\dfrac{95{,}000 \times 32}{\pi d^4} = \dfrac{2 \times 40.5}{d}$

rearranging we have $d^3 = \dfrac{95{,}500 \times 32}{\pi \times 81} = 12{,}008$

shaft diameter, $d = 22.9 \text{ mm}$

The kind of thinking involved is largely analytical and typically occurs when a problem is well defined and can be described mathematically. It is convergent because there is only one correct answer to the problem - it having well defined parameters and involving established conventions. In general these study materials will not deal with this type of design thinking or design problem.

ACTIVITY 12
Heat pump valve problem

So what kind of thinking will our creative techniques help to stimulate?

Let's look at another problem.

Let us suppose your firm manufactures heat pumps. You have been asked to create a device that will allow twice as much fluid to flow in one direction, when the pump is in the cooling cycle, as in the other, when the pump is in the heating cycle.

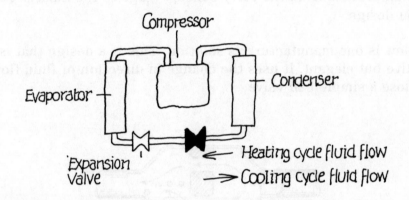

Heat pump valve problem

DESIGN PROBLEM REQUIRING CREATIVE THINKING

Suggest a design for a valve (▶◀) which will allow twice as much fluid to be pumped in one direction as the other.

Concentrate on the core problem ...... twice as much fluid to be pumped in one direction as the other. Make notes and sketch some ideas for tackling this problem in your Work File.

When you have done this, turn over the page to the Discussion of Activity 12 where you will see one particularly clever solution.

DISCUSSION OF ACTIVITY 12
Heat pump valve problem

Did you come up with anything?

At first I thought of various mechanisms for monitoring the rate and direction of flow, coupled to a device which opens and closes a simple valve to stop part of the flow in the heating cycle.

You may have thought of something different. Clearly there is no correct answer that can be calculated using a formula or rule of thumb. Tackling a problem like this requires creative thinking. It may be needed at any part of the design process from the early concept stage to the most detailed component design.

Shown below is one manufacturer's solution. This is a design that is not only creative but elegant. It uses the change in direction of fluid flow to open or close a simple ball valve.

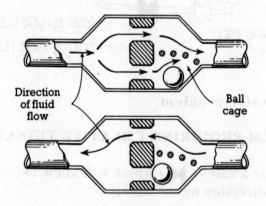

Heat pump valve solution
(Source: Middendorf, W. H. *Engineering design*, Allyn and Bacon; Boston, 1969)

The study materials will not necessarily give you the skills to come up with such elegant solutions. But our creative techniques will help you tackle problems requiring creative thinking - especially if you are having difficulties producing a solution using simply experience or more conventional methods.

ACTIVITY 13
Lifting mechanism problem

A third kind of thinking often required in design is concerned with *choosing between alternative solutions* to a design problem. You have to decide which is best according to certain criteria ... such as reliability and cost. Or you may have to select between different materials or components, trading-off one advantage, such as weight, for another, such as durability.

For example, suppose you had to choose between various mechanisms for raising and lowering a surface with a drive motor to reject items from a conveyor. The table is raised and returned with each revolution of the drive motor. Two possible solutions to this mechanism problem are shown below. How would you choose between them?

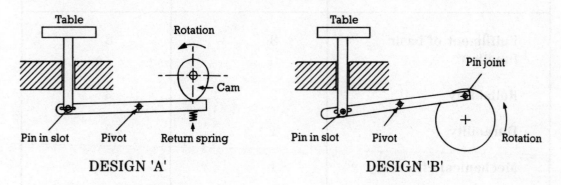

Lifting mechanism problem
(Adapted from: Hawkes,B. and Abinett,R. *The engineering design process*, Pitman: London, 1984)

DESIGN PROBLEM REQUIRING DECISION-MAKING

First you need some criteria against which to evaluate the solutions. I suggest the following evaluation criteria (or design objectives):

- Fulfilment of basic function
- Durability
- Cost
- Reliability
- Mechanical efficiency

Using scoring of:
3 = good solution, 2 = satisfactory solution, 1 = poor solution

Have a go at scoring design 'A' and design 'B' against the evaluation criteria given above. Do this before looking at the scores I gave which are listed in the table given in the Discussion of Activity 13 over the page.

DISCUSSION OF ACTIVITY 13
Lifting mechanism problem

How do your two mechanisms score on these criteria?

Well, both are good at performing the basic function. But design 'A' is likely to be less durable than 'B' because of the wear on the cam. So I rate 'A' as 1 and 'B' as 2 on durability. Going through each of the criteria in turn and rating each of the two designs will help you converge towards an optimum (or at least a satisfactory) solution.

Evaluation criteria (or design objectives)	Design A	Design B
Fulfilment of basic function	3	3
Reliability	2	3
Durability	1	2
Mechanical efficiency	1	2
Cost	2	3
etc.		

3 = good solution, 2 = satisfactory solution, 1 = poor solution

Mechanism Evaluation

Later in the Study Text, techniques are discussed which enable you to do this design decision-making in a systematic way.

ACTIVITY 14
Transport problem

Finally, there's a fourth kind of thinking required in design that is both *divergent* and *convergent*.

Very often it is necessary to define and analyse a problem before generating solutions or choosing between alternatives.

Suppose for example, you were given the brief of designing a monorail system for transporting people between a railway and an airport.

This brief and some solutions are shown below:

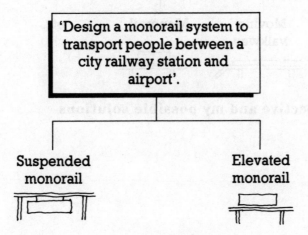

Objective as stated in the original brief

But does the brief describe the problem that the client really wants solved? Maybe it should be restated in more general terms as shown below? Enter your ideas for solutions in your Work File before looking at my suggestions on the next page.

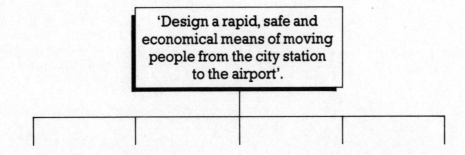

Objective restated in more general terms

DISCUSSION OF ACTIVITY 14
Transport problem

Restating the brief in more general terms immediately allows a variety of different solutions to be considered - a bus service, a moving walkway, etc. These are shown below:

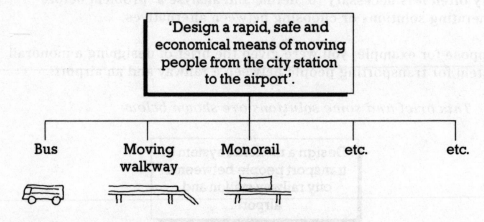

Restated objective and my possible solutions

DISCUSSION OF 'TYPES OF DESIGN PROBLEM'

Generating a variety of solutions may involve the use of creative techniques such as Brainstorming. But it's equally important to analyse a problem in order to be clear about the performance and the objectives of the users. Later in the Study Text you will find techniques (such as Objectives Tree and Performance Specification) which enable you to systematically define a problem during the initial stages of design.

So to recap, solving design problems may require us to use many *styles of thinking* - some divergent and some convergent. The study materials include techniques to help us define our design problems and specify the requirements of satisfactory solutions. They also include techniques for generating alternative solutions. Finally, we offer some techniques for systematically choosing between these solutions. What are not covered are techniques for carrying-out engineering analysis when tackling detailed problems.

I have summarised the different 'styles of thinking' and the problem-solving 'creative techniques' in the table below:

Style of Thinking (Design Skill)	Creative Techniques
Problem definition / customer requirements (specifying the objectives of a design project and criteria for evaluating alternative solutions)	Objectives Tree Performance Specification
Creative / inventive (creating a new design concept or a solution to a design problem)	Boundary Shifting Brainstorming New Combinations Analogies Morphological Analysis Checklist
Decision-making / evaluation (choosing between alternative solutions and improving existing designs)	Objectives Tree Performance Specification
Engineering analysis (e.g. component stress analysis)	Not covered in these materials

WORK TASK 1

Your Design Project

This is the first of ten Work Tasks.

The remaining Work Tasks are intended to give you the opportunity of trying the various creative techniques on a real design problem of relevance to your own situation.

So finding a suitable design project, or problem on which to apply the techniques, is important. That's the primary aim of Work Task 1.

You'll need a project that will allow you to apply the various techniques individually at first and then, as you become familiar with them, in combination with each other *and* with your usual methods of working.

I suggest that you scan through Section 2 to gain a general impression about what the techniques do.

They are not mathematical or highly technical, nor are they related to any specialist area of engineering. They are general methods for helping to analyse customer requirements, for generating alternative solutions and for choosing between those concepts according to given criteria. These are activities which tend in the main to occur at the earlier stages of design.

You will therefore need a project or problem that is not so specialised or complex an area that design features are rigidly determined by technical requirements. Nor should it be in an area in which customer needs and design solutions are so well-established that new concepts seem to be impossible. Although it should be said, that even in such areas, innovative designs can emerge as a result of applying the techniques.

The project should start with the initial stages of problem exploration and conceptual design and end with the early stages of embodiment design.

It is unlikely that you will get to the stage of detailed design using the techniques, although they are suited to tackling individual conceptual tasks in the detail design phase. You may of course wish to apply your usual approaches to analysis and detailed design - if you want to proceed beyond the embodiment stage.

Below I have listed some criteria to help choose a suitable project.

Criteria for suitable project topics / problem areas
<ul><li>There is a potential for new design concepts / solutions</li><li>Design solutions are not completely defined by scientific or technical constraints</li><li>Project involves stages of initial problem exploration and conceptual design (not just engineering analysis or detailing)</li></ul>

Below and over the page are some suitable project topics. These ideas are just to give you the flavour of suitable projects. They are not meant as topics for you to tackle, as they are too loosely defined for that. I hope they will help you, together with your tutor and/or manager, to identify an appropriate and relevant project.

Examples of the type of problem or project topic suited to the creative technique/s to be used
<ul><li>Investigate the feasibility of producing a range of garden accessories (e.g. a lawn rake) to be driven by a commercially available petrol-driven lawnmower. Produce a list of possible accessories, plus a set of initial drawings of the attachment part on the lawnmower and the mating connections on the various accessories. (For more details see Engineering Design Teaching Aids Programme EDTAP DE-6, The Design Council.)</li><li>Assume you are a project engineer for a firm that produces new and unusual electronic and electromechanical devices to be made in quantity by other firms. A customer requests that you design a door lock, to retail for £75 to £100, that recognises and unlocks for each of a chosen number of people using some means of identification (e.g. fingerprints, appearance, a magnetic card, etc.). Generate various possible design concepts and choose one to propose to the customer. (For more details see Algar,J.R and Hays,C.V *Creative synthesis in design*, Prentice Hall, Englewood Cliffs 1964.)</li></ul>

> *Examples of the type of problem or project topic suited to the creative technique/s to be used*

- Devise an improved car washing machine which avoids the disadvantages of conventional rotating brushes. High velocity water jets have not so far proved very satisfactory. (From Spotts, 1968.)

- Devise a low cost system to enable car owners to check the parallel alignment of the front wheels of a car at home. (From Middendorf, 1961.)

- Sketch some design proposals for a low-cost powered robot device for transferring materials from a storage point to a nearby automatic lathe. (From Hawkes and Abinett, 1984.)

- Devise a powered or hand-operated coin sorting machine to sort a mixture of coins in a hopper into denominations, for example for use in a one-man operated bus. (From EDTAP DE-3, The Design Council.)

- Investigate the performance requirements for, and provide sketch designs of, a system to enable the windows of a tall building, or the upper floors of a house, to be cleaned conveniently from the inside.

FURTHER SOURCES OF IDEAS FOR DESIGN PROJECTS

- A useful source of engineering design exercises and projects is the Engineering Design Teaching Aids Programme (EDTAP) produced by the Education Department of the Design Council, 28 Haymarket, London SW1Y 4SU (071 839 8000). EDTAP Booklets are provided as supplements to the magazine *Engineering Design Education and Training*, published three times a year by The Design Council.

- A book containing numerous engineering design projects is: Spotts, M.F *Design engineering projects*, Prentice Hall, Englewood Cliffs, 1968 (Chapter 2 especially).

- Other useful sources of exercises and projects include: Middendorf, W.H *What every engineer should know about inventing*, Marcel Dekker, New York / Basel 1981; and B Hawkes and R Abinett, *The engineering design process*, Pitman, London, 1984.

SECTION 2

Creative Design Techniques

STUDY GUIDE

From working through Section 1 you may have been able to classify yourself as a convergent or divergent thinker, or as more of a holist than a serialist. I hope that it also gave you an insight into your own powers of both 'creative thinking' and 'methodical thinking'.

I hope we share the same experience that *both* creative thought and methodical thought are necessary in the design process. The more methodical or systematic ways of working in design will also be covered. This section concentrates on helping you apply your creative thinking abilities in more powerful ways.

Firstly, the section takes a closer look at the creative problem-solving process. It draws on examples from science and engineering. Then it goes on to show you some techniques that can help in the creative process of seeking solutions. Each technique includes some examples of its use and is followed by a Work Task which enables you to apply the technique to your own product or design problem.

OBJECTIVES

After completing the study materials in this section you will be able to:

1. Describe some common features of the creative problem-solving process.

2. Apply the following creative techniques...

- Boundary shifting
- Checklist

- Brainstorming
- Objectives tree

- New combinations
- Performance specification

- Analogies
- Morphological analysis

..... to your own design problems.

What is the Creative Process?

I propose to start by looking in some depth at the creative process. You probably have your own thoughts about the creative process and how it works for you. For me, it is something to do with imagination; flair; insight; new perceptions of old problems; inventiveness; and determination.

I wonder how your thoughts compare with mine? And how they compare with some of the many other people who have also tried to describe creativity and the creative process. Here are some descriptions by various experts.

> 'Creativity is essentially a process of making new combinations of known pieces of knowledge ... a new idea is not imagined, it is produced by synthesis or at least analogy with known facts'.

> *(Farrendane)*

> 'Design can only be carried out by analogy'.

> *(Newman)*

> '... the creative act consists of a new synthesis of previously unconnected matrices of thought'.

> *(Koestler)*

> 'The creative process works by a process of analogy with the familiar'.

> *(Abel)*

These statements seem to have some accord with each other. There is also a general agreement about how the creative process operates, as these next statements suggest:

> '... the main principle in dealing with complicated problems is to transform them into simple ones. This recoding, or restructuring process depends upon bringing crucial aspects to the fore'.

> *(Jones)*

'A new and fertile pattern of thought may come from a conceptual realising of the universe into fresh classes and the making of new combinations of them. A good insight is likely to recognise the universal in the particular and in the strange ...'

(Hartmann)

This last statement is echoed in Koestler's writing where he describes the essence of creativity as:

'... unearthing the hidden analogies which connect the unknown with the familiar and show the familiar in an unexpected light' ... thereby making '... the strange familiar and the familiar strange'.

(Koestler)

Of course, this kind of definitional game can be played in many ways and in trying to define a word like 'creativity' there is a feeling of catching at mercury - once grasped , it breaks into a myriad of shining blobs.

So where does it take us?

There are several key characteristics of the creative process which emerge from these statements, such as analogy, restructuring, new combinations of existing material; new perceptions; seeing things simply and detecting universal principles. But these definitions and descriptions all tend to be rather abstract. At a less general level you may suggest other characteristics more specific to creative design practices such as:

- looking for crucial pieces of information as feedback, from a design or prototype

- skilful manipulation of a drawing or model which allows a rapid turnover of ideas

Let me suggest an activity which will help show the creative process at work.

ACTIVITY 15
Beetle problem

Our next more extended example is drawn from science , but it is about design - the design of a scientific experiment. More importantly it makes the general point that many design problems emerge as pairs of elements in seeming opposition:

- make it stronger **v** make it lighter

- make it perform better **v** make it cheaper

- make it more attractive **v** make it more robust

- make it do more **v** make it simpler to make

The polarities can be seen as a competition between equally valid demands. In this problem the polarity is:

- make it move **v** make it stay in the same place

Incidentally, this is the same polarity which faced the engineer Ove Arup in designing a footbridge. Try to remember this when reading the case study which follows this Activity.

So how do designers begin to reconcile these opposing demands?

Consider the following problem -

A team of scientists is engaged on experiments on the visual perception of beetles. They know the perceptual apparatus of the beetles is quite primitive, but that a beetle will respond to moving bands of black and white stripes. Of necessity, the stripes cover the interior of a cylinder. The width of the stripes and the rotation of the cylinder can be varied - so giving two measurable variables.

The beetle also responds in several measurable ways, by moving straight ahead, turning left or turning right or just stopping. However, it is clumsy and difficult to follow the path of the beetle with a suspended rotating cylinder!

For the purposes of the input stimuli the beetle is required to stay at the centre of the cylinder. For the purposes of the output response the beetle is required to move forward, turn right or left.

These then are apparently irreconcilable requirements - the beetle has to stay still and move at the same time.

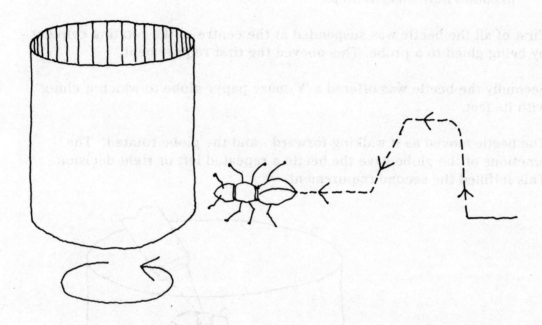

Irreconcilables: rotating cylinder and pathway of beetle

Can you devise a way of solving this problem? Spend 10 minutes thinking about it and put any ideas you have in the form of notes or sketches in your Work File.

DISCUSSION OF ACTIVITY 15
Beetle problem

The solution that was actually devised by the scientists with this problem had several steps:

First of all the beetle was suspended at the centre of the rotating cylinder - by being glued to a probe. This obeyed the first requirement.

Secondly the beetle was offered a 'Y'-maze paper globe to which it clung with its feet.

The beetle moved as if walking forward - and the globe rotated. The junctions of the globe gave the beetle a repeated left or right decision. This fulfilled the second requirement.

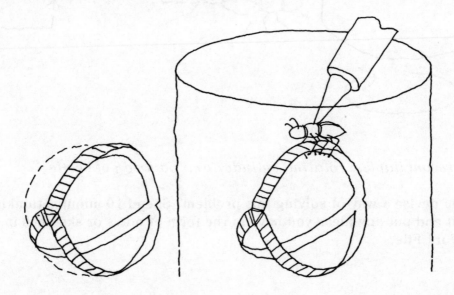

Reconciliation: beetle suspended at centre of cylinder, clinging to the 'Y'-maze globe.

This was a solution of great simplicity and great ingenuity - but in testing it the scientists discovered a critical factor which they had overlooked. The beetle would not hang onto the globe.

By trial and error, the experimenters found that the weight of the globe was the critical factor. Perhaps you can guess that critical weight?

Globes which were too heavy were dropped and globes which were too light were dropped. They discovered that the critical weight of the globe *equals* the weight of the beetle.

In retrospect this made sense. The beetle in its natural environment was accustomed to walking on vertical surfaces and the underside of leaves and twigs. The appropriate muscular tension in its legs therefore was prompted by its own weight.

(*Source:* Dean Woolridge, *Machinery of the Brain*, McGraw Hill)

LEARNING POINT This case illustrates not only how two directly opposed requirements were brought into a successful fusion: but how the solution depended on a critical factor which was discovered only through testing - a factor that it was extremely difficult to foresee.

Many other problems - especially in design - display the characteristics of having seemingly contradictory requirements. Over the page you will find an example from structural engineering.

This footbridge at Kingsgate, Durham, is the subject of the following Case Study.

CASE STUDY: EXPANSION JOINT

Some years ago, the engineer Ove Arup was working on the design of a pedestrian footbridge for the University of Durham. The bridge consisted of two channel section beams each supported on four struts rising from two bases. The construction was insitu reinforced concrete.

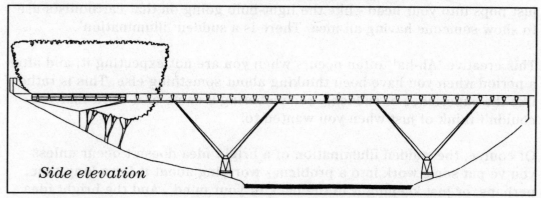

Side elevation

The two identical halves of the bridge needed to be pinned together at the centre. The joint had to be capable of transmitting small vertical forces from one half to the other, yet that joint also had to allow for thermal expansion. Thus the contradictory requirements were that the bridge must be pinned in place, but it must be allowed to move. Reproduced here are some sketches from Ove Arup's notebook. As you can see he is exploring a large scale joint (of phosphor bronze) which allows movement in one direction but not the other.

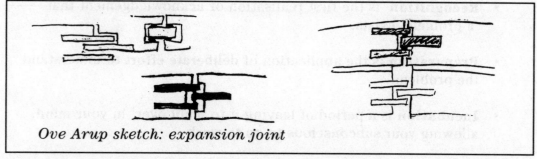

Ove Arup sketch: expansion joint

And here is the actual design solution that was finally developed.

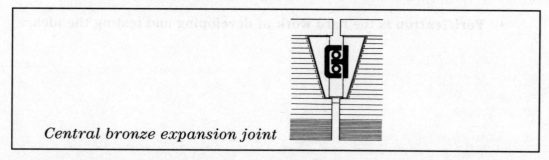

Central bronze expansion joint

The ingenuity of this design lies in resolving the contradiction, and in doing it with a form which is neat and legible. By that I mean the possible movement of the bridge can be read by looking at the completed joint.

IS THERE A PATTERN TO THE CREATIVE PROCESS?

Psychologists and others have studied accounts of creative thinking by a wide range of scientists, artists and designers. As you may have experienced yourself, most people report that there is often a sudden creative insight that suggests a solution to the problem. The brilliant idea just pops into your head - like the light-bulb going on that cartoonists use to show someone having an idea. There is a sudden 'illumination'.

This creative 'Ah-ha!' often occurs when you are not expecting it, and after a period when you have been thinking about something else. This is rather like the experience of suddenly remembering a name or a word that you couldn't think of just when you wanted to!

Of course, the sudden illumination of a bright idea doesn't occur unless you've put some work into a problem - worrying about it for a long time, perhaps, or just keeping it in the back of your mind. And the bright idea itself is usually just the germ of an idea that needs a lot of further work to develop it into a proper, complete solution.

This kind of thought sequence occurs often enough in creative thinking to suggest that there is a general pattern to it, like this:

Recognition- Preparation- Incubation- Illumination - Verification

- **Recognition** is the first realisation or acknowledgement that 'a problem' exists.

- **Preparation** is the application of deliberate effort to understand the problem.

- **Incubation** is a period of leaving it to 'mull over' in your mind, allowing your subconscious to go to work.

- **Illumination** is the (often sudden) formulation of the key idea.

- **Verification** is the hard work of developing and testing the idea.

FUN OR FRUSTRATION?

Creativity is usually presented as 'fun' - in group exercises, especially, the emphasis is often on lightheartedness and humour. There are certainly strong parallels between creativity and humour - they probably both draw on the same kind of thinking.

But the driving force behind creative thinking in design can also be 'frustration' - a sense of dissatisfaction with things the way they are, a feeling that almost everything could be improved upon! So the fun and frustration of creativity probably go together - they are two sides of the same coin.

Look around you at the products you come into contact with - not only at work but in your daily life. Aren't all of them inadequate in some way? Couldn't you devise improvements?

Creative thinking isn't something that you just 'switch on' when necessary. It's an attitude of mind that is always there. And one way to cultivate it is to frequently spend time thinking about everyday objects - about the frustrating things that surely *ought* to be better!

Here are a few examples that occurred to me of actual 'needs' where some creative design thinking is needed:

- a way of holding the telephone to my ear whilst using both hands (e.g. when writing messages)

- a bicycle lock that's less cumbersome than chains and padlocks

- a chair that 'grows' with a child

- a better way of joining sheets of paper than staples

- a quick, cheap way of taking local aerial photographs.

Many of the creativity techniques that follow can help in generating solutions to such problems. But we don't have to rely on magic 'techniques', and we don't have to wait for problems to be 'set' for us - they are all around us!

Using Creative Techniques in Design

Some of the techniques that I will be presenting are ones with which you may already be familiar. Perhaps you use versions of some of them already. My purpose in this section is to help you to make your use of creative techniques more open and explicit in the design process. This provides the benefit of not only making your own pathway of decision-making clearer to you, but also of allowing other members of the design team to contribute to the decision-making process.

The techniques with which we deal in this section are ones that *open-up possibilities* that might otherwise be neglected. The techniques covered are:

- ① **Boundary Shifting**
- ② **Brainstorming**
- ③ **New Combinations**
- ④ **Analogies**
- ⑤ **Checklist** *find out from PowerPoint, word*
- ⑥ **Objectives Tree** *find out from PowerPoint or word.*
- ⑦ **Performance Specification**
- ⑧ **Morphological Analysis**

Boundary Shifting

This is a technique that no doubt you use already - if under other names, or perhaps without realising it! The description and explanation of it here are intended to make you more conscious of it as an explicit technique - and perhaps apply it to situations of your own choosing.

Normally boundary shifting is introduced from outside the design process:

'Here is a new material that I'd like you to try'

or perhaps more typically:

'We need to cut our manufacturing costs by 10 per cent - can you do a redesign?'

No doubt you have a general sense of such boundaries or constraints. When starting a design you are clearly limited by cost, materials, structure, geometric possibilities, law, manufacturing processes, performance characteristics, and so on. But as you know, each constraint does *not* operate independently. The change of one constraint has repercussions in all the others - as if they are pinned together in an enclosing net of linkages.

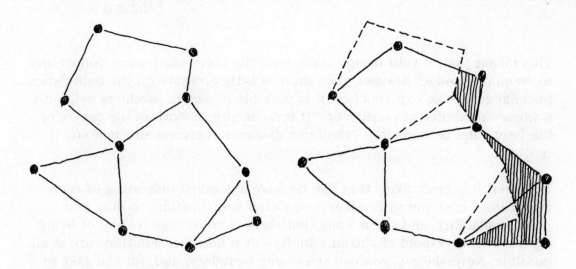

Constraints 'pinned' together

New areas of possibility enclosed by shifted constraints

I find it convenient to think of these constraints being laid over a total field of possible solutions - with the impossible and improbable designs falling outside the boundaries; the exploratory and risky solutions within but near the boundaries; and conventional known solutions in the centre of the field.

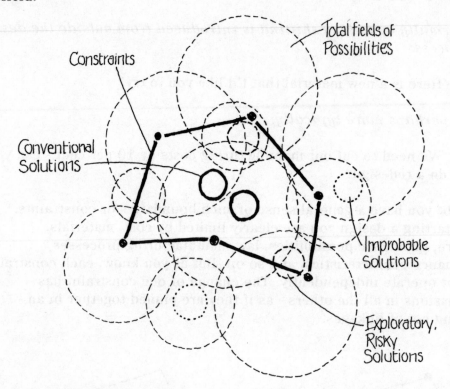

This means that, if your design tasks typically are evolutionary 'follow-ups' to standard product designs, then there is little pressure on the boundaries for change. It also explains how it is possible to design products with only a vague knowledge of constraints. It feels secure to work in the centre of the field - and perhaps it is - until you discover everyone else is doing it too!

However, it is more likely that you do have a detailed indication of each constraint (cost and price targets, material specification, tooling and plant availability, and so on) and that there is an intense feeling of being right on the threshold of absolute limits - as if no other solutions are at all possible. Nevertheless, you can select one boundary, and, for the sake of further explorations, change it. This will then simultaneously open up a new range of possibilities and close down others. Some of these new possibilities may prove more plausible than you think!

Because this technique deals in sets of solutions and fundamental constraints it is best applied at the *early* stages of the design process. It is particularly appropriate for testing out the assumptions of a design 'brief'.

60

USING THE BOUNDARY SHIFTING TECHNIQUE

CASE STUDY : FOOTBRIDGE

The first brief for a footbridge by the University of Durham to Ove Arup was for a short span, low level bridge. The line of this bridge was chosen to pass over a minimum gap, and therefore, should provide the cheapest solution. Thus it appeared that the cost constraint implied a particular, rather mediocre, solution.

However, Ove Arup was able to argue for a higher level bridge on a different line of approach. This removed the necessity for a flight of steps on either side which would have been a great inconvenience to the users of the bridge - as well as an extra cost beyond the structure of the bridge.

By an ingenious, delicate structure the engineering designers were able to keep the costs within the original target and provide a more elegant design - in keeping with the surroundings.

In this case the design brief was challenged, the boundary of the problem (literally) extended, while the major constraint of cost remained fixed. The change in line of the bridge stimulated a whole new set of possible geometries and constructions.

Note: This footbridge is the same as that referred to earlier in the *Expansion Joint* case study.

You may wish to look again at the drawing of the side elevation which shows clearly the geometry of the completed bridge (see page 55).

(*Source:* O.Arup, 'Kingsgate Footbridge', *Northern Architect*, March, 1966)

CASE STUDY: HAIRDRYER

Charles Mauro, an American industrial designer specialising in the ergonomic aspects of design, conducted a careful examination of the user requirements for hairdryers. He concluded that many of the dangers and problems of hairdryers derived from the electric motor and fan, i.e. their weight to the hand, the electric source near wet material, their storage when not in use.

His prototype solution proposed a wall mounted power unit and fan with a long flexible hose.

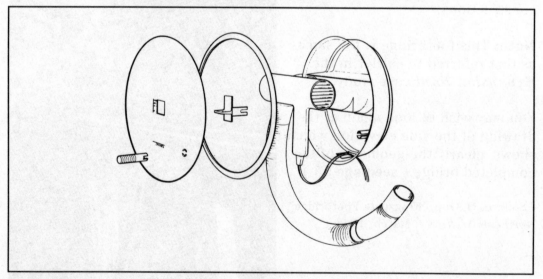

In this case the conventional solutions are challenged, the boundaries of human use and safety are pushed back, while other considerations are given less importance.

(*Source:* C.Mauro, 'Can hairdryers be safer? Research says Yes', *Industrial Design*, May/June, 1978.)

WORK TASK 2

Creating new solution concepts by boundary shifting

The intention of this Work Task - in common with other Work Tasks in this section - is to be provocative in suggesting new concepts for the redesign of an existing product, or for a completely new product. (If you have not obtained the necessary information required by Work Task 1, then you will need to discuss this with your manager and / or tutor.)

1. Write down a list of the major constraints that are acting upon the design of your product. (Note: include those factors which would have been given as an original design brief'.)

2. Underline those that are well supported by documentation.

    e.g.- target costs
       - standard components.

3. Try to add constraints to the list which are implied, but not well documented,

    e.g.- using existing manufacturing equipment
       - pleasing a well established customer
       - using a familiar supplier

4. Add others which you know about intuitively and by experience,

    e.g.- competitors' successes
       - market preferences

5. From this list of constraints choose one and modify it (preferably not costs) according to a new hypothetical circumstance

    e.g.- improved material

6. Think through the implications on all the other constraints.
    What new range of possibilities opens up?

7. Make sketch proposals for some of those new possibilities

The lists, notes and sketches should be entered into your Work File for detailed discussions with your manager and tutor. The possibility of your proposals being further developed will doubtless depend upon the strength of the arguments you are able to present.

Brainstorming

Brainstorming is a technique for opening up possibilities. It is appropriate to the early stages of conceptual design: that is, after the briefing process and before the principal form of the design has been established.

As a creative technique it is looser and more unpredictable than some others. It *is* a licence to fantasise and have daft ideas. The rational work of analysing the ideas comes *after* a *deliberate* suspension of critical judgements.

The justification for such a technique can be put as follows:

In the design process there is inevitably a high mortality of ideas. At each stage a great number of ideas is explored but only a few pass on to the next stage of detail. This converging process is represented sometimes as a logarithmic decline.

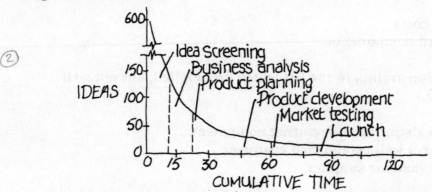

Thus the design process begins with an infinitely large number of shadowy ideas and is completed in one highly realised specification. As you know, the early parts of the process are usually characterised by a wide ranging, and hopefully exhaustive review of possibilities. If, instead, the early part of the process happens to be characterised by a fixation on one or two ideas, or worse still no ideas at all, then clearly there is cause for unease.

Brainstorming is a technique which generates ideas and promotes wide ranging explorations. You may perhaps return to the initial fixed ideas, but at least, in this new frame of mind, you can be more secure, and not have the lurking suspicion that you just *might* have overlooked some obvious solution. To avoid disappointment the main point to grasp is that this technique will give you a very high percentage of duff ideas. It is unlikely to offer you a competent complete solution on a plate.

What it *will* do is give you many new insights and a greater clarification of the problem. It is a great help to review a range of wrong answers, and know why they are wrong, in the steps towards getting better answers.

USING THE BRAINSTORMING TECHNIQUE

EXAMPLE: ROAD SAFETY

In order to get the hang of Brainstorming, work through this example with me. Spend 5 minutes jotting down as many ideas as you can in response to the question:

How can we improve road safety?

Here is my list, in no particular order:

1. Reduce speeds
2. Tramways, rails, magnetic lines
3. Individual small low powered cars
4. Cars banned from city centres
5. Knowledge workers at home, less commuting
6. Improved railway system
7. Inflatable cushion cars
8. Automatic braking
9. Electric cars
10. Electronic ordering of goods
11. All round transparent plastic for improved vision
12. Increased use of cheaper taxis
13. Sleeping, or awake, police officers at accident spots
14. Improved bus services
15. More child education in safety
16. Horror news spots
17. Speed governor to all road vehicles e.g. max. 70 mph
18. Pathways for cyclists
19. Life ban for convicted drunk drivers
20. Repeated driving tests every five years

This is a mixture of nutty and not so nutty ideas - with which to move to the second phase of Brainstorming, combining and improving on the ideas generated so far:

Inflatable cars of clear plastic (ideas 7 and 11) offer something - at least in visibility. They may burst in accidents rather than absorb impact. Balloon shapes are quite fashionable and aerodynamic!

Working at home, delivery of goods by rail, and electronic ordering would cut down shopping and commuter trips (ideas 5 + 6 + 10).

An automatic pilot is implied by some ideas (ideas 2 + 8 + 17) with strict controls on speed and route. The driver merely presses an ON/OFF button.

Perhaps we have generated three ideas worthy of pursuing! Not very brilliant ideas and all dependent on new high technology.

Other combinations seem more promising; cheap and plentiful public taxis, for instance (derived from ideas 12 + 14); or non-hardware methods, such as child cyclist education campaigns (ideas 15 + 16 + 18).

Other ideas attack specific causes of accidents and seek legal solutions ... drunk driving, excessive speed, and alterations to vehicles (ideas 1 + 17 + 19).

When you have settled on two or three combinations of ideas that you feel have some feasibility you can subject them to stricter criteria.

> *For example, try a quick evaluation of the idea from the point of view of a relevant party, such as:*

> a)   a town's road safety officer

> b)   the Department of Transport providing funds

> c)   a lady shopper who relies on her car

> d)   a busy travelling salesman.

Choose the interested parties in order to give criticisms and objections to the idea, as well as a new fund of information.

The generation of many ideas (here I have generated twenty ideas in ten minutes) leads naturally to an attempt to group them in clearly associated sets. This *classification* is itself an important part of clarifying the problem.

> *The solutions, above, seem to me to fall into three categories:*

> a)   hardware improvements

> b)   system restrictions

> c)   system enlargements.

Your own categories may be different but it should give you an outline of the main factors of the problem.

RULES OF BRAINSTORMING

Brainstorming is probably the best known of the techniques included in the various available books. For this reason, I am starting with Brainstorming and then will show you how it can be used in combination with other creative techniques.

Brainstorming is well known and popular because it is easy to understand. That doesn't mean it's as easy to apply as it may seem. But if you use it properly, brainstorming is a very quick and valuable method of generating lots of ideas .

It's usually best used at the early, conceptual stage of a design project and rather less likely to produce ideas for solving detailed technical problems; it's particularly useful as part of a divergent search for radical solutions. For example, a brainstorming session is likely to be better at thinking up possible applications for new components or materials than solving a detailed problem in electronic circuit design. However, given the right mixture of expertise, brainstorming has been successfully applied even to highly technical problems.

Brainstorming is sometimes seen as just a fancy name for having a group discussion or for throwing ideas around. It is of course to do with the generation and discussion of ideas, but Brainstorming proper has extra rules that help the ideas to flow.

Rule 1. No criticism is allowed at all

Ideas should not be squashed by someone saying, or even you thinking yourself, 'that won't work because ...' or 'that's silly' or 'it's been done before'. Brainstorming requires one to relax the usual inhibitions on what we say or think.

Rule 2. Try to produce as many ideas as possible

Keep a record of them for later evaluation and development.

Rule 3. Make it a group activity

Brainstorming is best performed in a group of four to eight people. But it can be difficult to get a Brainstorming session going because people often feel self-conscious about it.

I'm going to start with an *individual* brainstorm - so you have rehearsed it yourself before trying to persuade others to join in. To start a Brainstorming session it is important to pose a suitable question. If the question is too narrow it will limit the range of ideas. If too broad, it can lead to lots of irrelevant material.

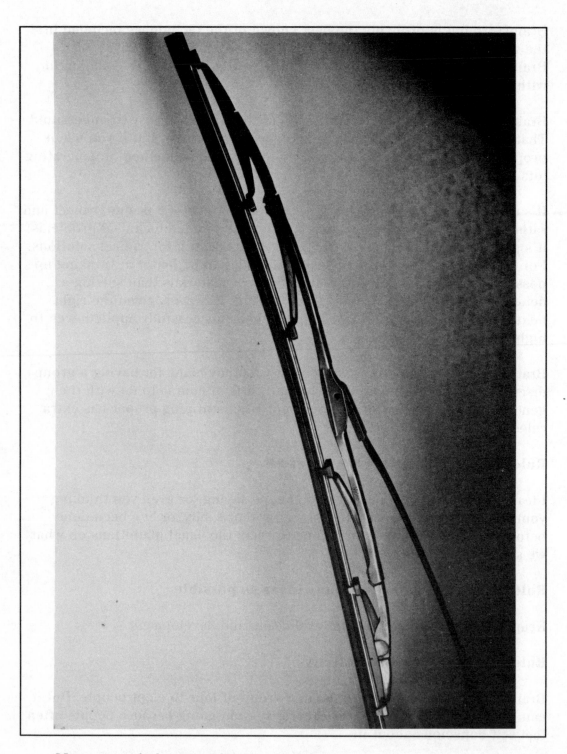

Motor car windscreen wiper

ACTIVITY 16a
Individual Brainstorming (Windscreen wiper)

Let us suppose you worked for a vehicle component firm and were given the problem of thinking up alternatives to conventional windscreen wipers on motor cars. Something that would avoid their deficiencies.

How would you phrase a question to start an individual brainstorming session on this topic?

Formulating a good question is not as easy as it might appear.

> *You want to avoid too narrow a question like:*

> 'How can we improve windscreen wipers?',

> *or too vague a one like:*

> 'How can we keep the windscreen clean and dry?'.

> *My suggested question is,*

> 'How can we eliminate the conventional windscreen wiper?'.

You may have come up with an equally good question.

Note your question in your Work File.

Now have a go at an individual brainstorm on the windscreen wiper problem. Take my starting question or your own, and spend about 5 minutes producing as many ideas as possible to tackle it.

Remember, do not evaluate the ideas at this stage. Just write down in your Work File at least ten ideas as fast as they come into your head. Your ideas will be used again later in Activity 16b.

When you have done this, compare your ideas with mine listed in the Discussion of Activity 16a on the following page.

DISCUSSION OF ACTIVITY 16a
Individual Brainstorming (Windscreen wiper)

How many ideas did you manage to produce?

My ideas are listed in the table below:

- Absorb water in the screen
- Rotating windscreen (like used on ships)
- Vibrate the screen
- Don't drive in the rain or snow
- Vehicles in tunnels
- Electrical charge to throw off raindrops
- Blow away the rain or snow
- Spray water over the screen
- Removable blade
- No windscreen / driver wears goggles
- Hinge the screen down in the rain
- Coat the screen with water repellent
- Wipe with a rag
- Drain away water
- Dry screen moves into place, wet screen moves away and is dried
- Blade moves in parallel tracks across the screen

One of the benefits of group brainstorming is that the ideas of one person trigger off further ideas in other participants. So I suggest that you next read through my list and then add to your own list any further ideas stimulated by mine.

We could stop there and group together both our sets of ideas on the windscreen wiper problem and begin to evaluate their potential for further development. But Brainstorming on its own does not necessarily generate the most original ideas for tackling a design problem. It is better used in combination with other creative techniques .....

New Combinations

You will have probably encountered ingenious objects that combine previously established separate *forms* in new ways. I rather like the 'flexmaster' screwdriver which combines three different forms to enable three functions to be performed. It has *attributes* which enable you to strip the insulation, twist the wire and connect domestic electrical connections, such as 13 amp plugs.

From this kind of common experience of artefacts, then, you already know that new forms can arise usefully from the recombination of existing forms. At a relatively unsophisticated level, most artefacts can be analysed as a set of simple forms or combination of attributes. The explicit articulation of these elements and their systematic recombination can give new and potentially useful designs.

The essence of the New Combinations technique is to look for new combinations of known elements. These elements may be described as geometric elements (i.e. FORM) or as performance features (i.e. ATTRIBUTES), or more crudely as whole object descriptions (i.e. an object capable of stripping, twisting and connecting electrical wires becomes a 'Flexmaster' screwdriver).

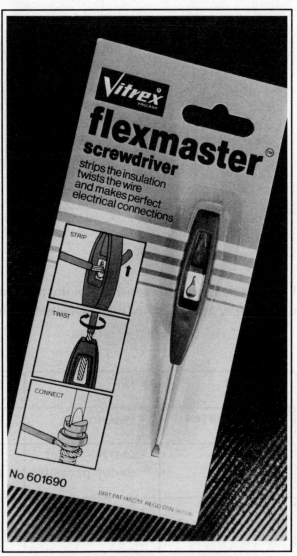

'Flexmaster' screwdriver
(Manufactured by Vitrex - reproduced with permission)

The outline of the New Combinations technique given here may be extended and amplified using other more systematic techniques given later in Section 2 (e.g. Morphological Analysis). The New Combinations technique can be used in a systematic or loosely associative way.

USING THE NEW COMBINATIONS TECHNIQUE

CASE STUDY: D-I-Y PLUMBING

I have been worrying about connecting hot and cold water taps to an existing plumbing system - for a washing machine. I do not find the traditional solutions satisfactory; they often require specialist expertise, tools and materials, draining the whole plumbing system, and so on.

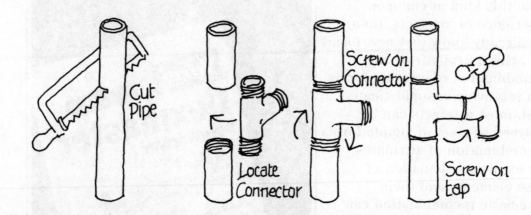

Traditional solution

I can analyse the conventional solution under headings of *form* and *attributes*, as suggested by the New Combinations technique.

Form	*Pipe*	*T-Connector*	*Tap*
Attributes	- copper	- copper	- plastic
	- carries water under pressure	- connected after pipe cut	- on/off valve
	- must be drained		

The nuisance to the amateur handyman is in draining the system, cutting the pipe exactly, and ensuring leak-proof joints.

Let me describe the process with the words:

PIPE - CUT - CONNECTOR - TAP - SEAL

My first thought was to combine Connector / Tap. This implied PIPE could combine with CUT - as a pre-drilled pipe. But this suggestion relies on a hypothetical series of partially drilled holes and a push fit connector, that may or may not need sealing.

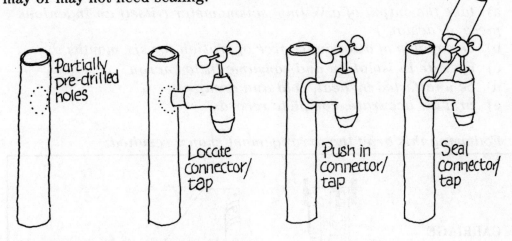

New variant (hypothetical)

The best solution would be a multiple combination:

CUTTER / CONNECTOR / SEAL / TAP

In other words, a screw-on tap which simultaneously drilled and sealed the supply pipe.

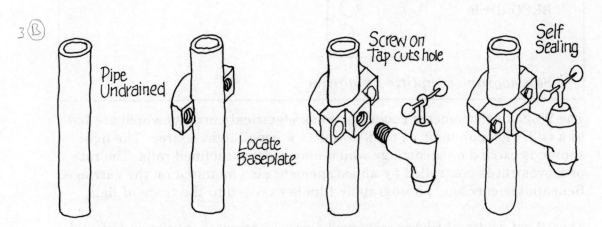

Combination: CUTTER / CONNECTOR / SEAL / TAP

But is such an idea workable? Well yes. As you may know, an Italian-made 'Self-boring Plumbing Kit' is marketed in the UK - which is exactly that combination of attributes! It is easy to fix and it is effective. Moreover, it does not not need the plumbing system to be drained.

CASE STUDY: SEISMOGRAPH

Christopher Spencer is a scientific instrument designer with a background in physics and optics. He was asked to design a machine which would record earthquake vibrations. His brief required that his design would:

a) *take the output of a Wilmot seismometer (itself an ingenious piece of design)*
b) *be capable of attendance-free operation for six months*
c) *operate in isolation and consume little energy*
d) *be unaffected by heat, cold and damp*
e) *provide accurate, complete records*

Following this brief the arrangement shown resulted:

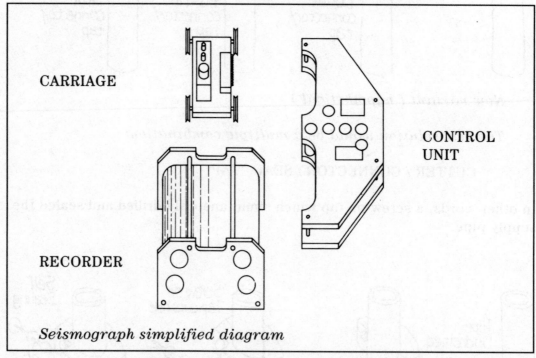

CARRIAGE

CONTROL UNIT

RECORDER

Seismograph simplified diagram

The Wilmot seismometer generates weak electrical currents which are fed to a coil. This coil induces oscillations in a pencil light source. The light source is carried on a carriage which moves down inclined rails. The rate of movement is controlled by an escapement also mounted on the carriage. Beneath the carriage, photographic film is exposed to the trace of light.

The 10 cm width of film is traversed every 15 minutes. At the end of each 15 minutes the carriage moves back to the top of the incline and the film advances a small stop. The spool contains enough film for a 6 month record. The winding-on of the film and the relocation of the carriage are powered by a battery. The seismograph has been used successfully in field locations, providing a satisfactory record when left unattended for six months.

These photographs show the working prototype:

Complete assembly showing inclined plane

Recorder showing photographic film

This was not a design achieved by systematic use of the New Combinations technique, but it does show the bringing together of familiar antecedents. For example, here is a list of artefacts brought together in this design:

- Camera using aerophotography film with automatic wind-on.
- Oscilloscope, used here to project light onto film rather than electrons onto a sensitised screen.
- Inclined plane allows the carriage to fall under its own weight. The controlling escapement is a standard item made by a Swiss manufacturer.

WORK TASK 3

Searching for new combinations

Consider carefully the product design problem that you have already
obtained in Work Task 1. See if you can generate some design improvements in the following way:

1. List the elements of the existing product under headings of
   FORM and ATTRIBUTES.

2. Identify combinations of elements that either are current sources
   of problems or dissatisfactions or seem to suggest potential new
   areas for improvements. (You may find it helpful to list
   systematically all the possible combinations in twos or threes, so
   that you don't overlook some unfamiliar, but perhaps promising,
   combinations.)

3. Make notes or sketches in your Work File for design
   improvements, redesigns, or possible new combinations of the
   existing elements.

These suggestions should be 'worked-up' sufficiently to form firm
proposals for discussion with your manager and / or tutor.

Analogies

Doubtless, you probably experience the need to range widely in your search for appropriate solutions to design problems. Sometimes your ability to do this relies upon detailed knowledge of a specialised field, but often it relies upon 'drawing-in' ideas from unexpected quarters. In this wide-ranging kind of search, the details of objects are less important than the principles they embody. You want to be able to scan quickly the associated sets of concepts or principles, so as to generate appropriate starting points for your new design.

When you settle upon an existing object as offering in principle a precedent - especially if this object is remote from the immediate problem - this is termed analogical thinking, or thinking through analogies.

In this frame of mind the insight that you have is to say 'This is like that!'. For instance, a concrete strut is like a rhubarb stem, or a housing layout is like the pattern on a doormat (see following examples).

Thinking analogically is, of course, a characteristic of creative minds outside design and engineering, and is used especially by poets:

'All the world's a stage ...'

William Shakespeare

'If this country were a sea ...'

Ted Hughes

'In full moon shiny white shoe-polish bright light gleaming like picture book paper ...'

Michael Horovitz

So the analogies you call upon during the 'thinking' process, originate from that part of the mind which deals normally in creative associations and metaphors. There are no sure-fire methods for promoting this ability - although the guidelines that follow in Work Task 4 may help.

Your ability to 'draw-out' useful analogies will depend less on those guidelines than upon:

- an openness to the possible contribution of disparate and accidental ideas

- an ability to step back from the overwhelming detail of a problem

- a good knowledge of the principles of a wide variety of devices

The success of this technique depends more on the repertoire of ideas you can 'draw-out' than upon rules. Perhaps you already like to keep a mental filing cabinet of ingenious devices and interesting ideas - storing them away for future use. You may like to cultivate that store of ideas more self-consciously and explicitly - in a notebook, a real filing cabinet or whatever you are using as your Work File.

USING THE ANALOGIES TECHNIQUE

CASE STUDY: STRUCTURAL SUPPORT / RHUBARB STEM

The analogy between structures and living plants is very often made. Nature provides a repertoire of ingenious forms that we may borrow.

For instance, in designing the Kingsgate Footbridge, Ove Arup likened the main structural members to a rhubarb plant. In early sketches of the design the concave stems of rhubarb were copied. Later versions made the struts convex.

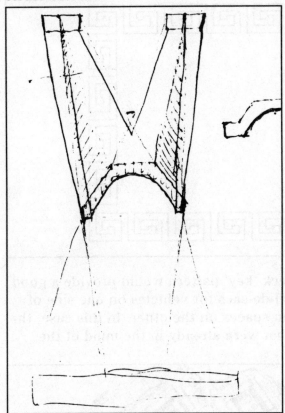

An early sketch of the structural supports

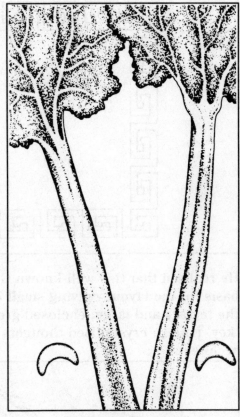

Rhubarb stems were used as analogies

In a later interview, Ove Arup said, 'The sketches show the idea I had first for the struts. They are a very funny shape. It was a question of whether they should be round, circular or half circular ... I saw them like a thing growing out, like a rhubarb stem. But after that I thought 'It's bad, water could collect in these things and you couldn't get rid of it'.

The analogy was a stimulus to how the overall form of the bridge should be constructed, and to the half-round section of the struts.

In detail the direct copying of the original did not work. This is usually the way with such analogies - they provide a diagrammatic idea - rather than detailed solutions.

CASE STUDY: HOUSING LAYOUT / DOORMAT PATTERN

The architect Richard MacCormac was working on a large housing scheme for the London Borough of Merton. It required pedestrian spaces on one side of a terrace and vehicle access on the other. The terrace was to take a zig-zag path to provide different varieties of enclosures. However, the geometry of the layout was proving stubborn.

Then the pattern on his doormat made its contribution:

He realised that this well-known Greek 'key' pattern would provide a good basis for the layout - giving small cul-de-sacs for vehicles on one side of the terrace and large, enclosed green spaces on the other. In this case, the 'key' pattern crystallised thoughts that were already in the mind of the architect.

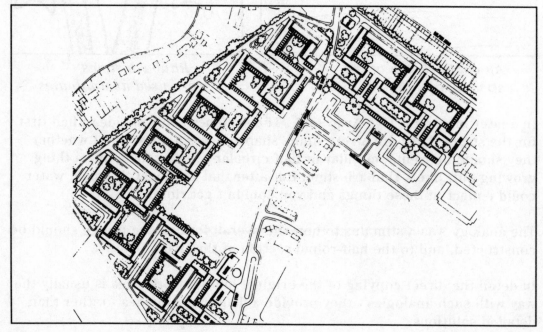

ACTIVITY 16b
Analogies (Windscreen wiper)

The aim is to think of things or situations that are similar to the problem you are facing.

One kind of analogy you probably already use when designing is to think of other devices or systems that have a similar function. In my brainstorm on the windscreen wiper problem (Activity 16a) for example, I remembered the rotating screen system sometimes found on ships. So first try to think of other devices for removing water or snow from surfaces, or other situations where a similar problem arises. Write your ideas in your Work File. When you have done this read on.

Another very useful stimulus for creative design is to look for analogies in nature - in plants and animals and so on. Remember the case study example of a rhubarb stem providing the idea for the structure of the main supports of a bridge ! Another example is a novel optical data storage system which was inspired by the way in which the eye of a moth absorbs light.

So, next try to think of any ideas for approaching the windscreen wiper problem offered by nature. Perhaps fish or water plants might offer ideas. Note your ideas in your Work File.

Looking for solutions in other man-made objects or in nature are called 'direct' analogies. Another kind of analogy is the 'personal' analogy.

Here you try to imagine you were inside or even part of the object being designed. How would it feel for example, if you were the windscreen being hit by raindrops? To keep dry perhaps you might try to repel them in some way or perhaps absorb them.

This kind of imagining may seem very fanciful, but it can give you a completely new perspective on a problem and hence useful ideas for solving it. So try to envisage yourself as the windscreen trying to keep dry or as a wiper blade moving across a screen, or even as a rain drop about to hit the screen and being prevented somehow. Try this personal analogy method now. Note down any resultant ideas in your Work File.

DISCUSSION OF ACTIVITY 16b
Analogies (Windscreen wiper)

I hope these various analogies stimulate some good ideas. My list of ideas is given below. Several of the ideas are similar to ones of which I had already thought. There are also some new ones, like using a recently developed chemical coating that prevents rain from forming into drops and hence allows water to run off without obscuring vision.

Direct Analogies

(a) Similar situations / devices / solutions in the man-made world:

- Rear windscreen has heater to remove condensation.
  (Use a heater to evaporate rain ?)
- Rain on spectacles. (Apply a non-mist surface to windscreen like anti-mist cleaning materials used for spectacles ?)

(b) Similar situations / solutions in nature:

- Water runs off a duck's back. (Coating on the screen that allows droplets to run off quickly. A dirt-repellent chemical coating called Micatex which prevents rain from forming droplets was demonstrated on BBC TV's *Tomorrow's World* in January 1986.)
- Eyelids blink water away.
  (Some kind of self-wiping eyelid built into the screen ?)
- Dog shakes off water.
  (Vibrate the screen to throw off the rain or snow ?)

Personal Analogies

- What is it like to be the windscreen trying to keep dry in the rain ? - shake off water; blow it away; absorb water; repel drops; burn away drops.
- What is it like to be a wiper blade removing rain from the screen ? - absorbent wiper blade would only need to sweep clean occasionally.

WORK TASK 4

Analogical thinking

Try to apply the technique of analogical thinking to your product's design or a product with which you are very familiar.

Think first about the product overall - think about its general FORM and its principal FUNCTION.

> *Ask yourself:*
>
> 'What is it like?'
>
> 'What is it doing?'

Then, if it is a fairly complex product, think about some of its key components, their forms and functions.

> *In an open-minded, 'brainstorming' kind of thinking (you can use a group to help with this technique, too!), try to generate several analogies in each of the following categories:*

- *Direct analogies* - perhaps taken from biological examples,both plants and animals. 'How does this occur in the Natural World?' Or, you can use direct analogies with other kinds of machines or products.

- *Personal analogies* - in which you imagine what you would 'feel' like if you were the object. 'How would I be coping with the job?'

- *Symbolic analogies* - which call for poetic or abstract metaphors. 'How might a poet describe this?'

- *Fantasy analogies* - in which you 'conjure-up' a fantastic or magical solution.

Like brainstorming, the idea is to generate lots of 'crazy' analogies, because in just one or two of them there might be the seeds of a good, original solution.

When you have generated some analogies, then have a look at each in turn and ask yourself if there are ways you can use the analogy in new design solutions or improvements. The best analogies should be 'worked-up' for discussion with your manager and / or tutor.

DISCUSSION OF WORK TASKS 3 & 4

The purpose of these Work Tasks is to encourage *free thinking*, but free thinking within a defined framework.

They should be approached in a relaxed exploratory mood.

Like any other exploration you do not know exactly where you are going and you do not always know the right process. These things emerge as you work on the problem.

At the conclusion you should reach a solution of sorts and have a strong sense of the journey. It helps if you keep some kind of narrative record of that process, so that you can subject the process to some kind of retrospective examination.

Here is some general advice to assist you:

- Take one idea at a time

- Accept it uncritically on its own terms

- Keep good drawings and notes

- When stuck go back through your notes for a new starting point

- Make an overall map of your journey, note which points are cul-de-sacs, which parts seem easy and at which points you were lost in a maze.

You will find that the process is partly analytical, partly synthetic, partly diverging and partly converging. There are some moments of logical construction and some leaps into space. Sometimes there is a diligent search for connected ideas, sometimes a flash of intuition which binds everything together.

Most of all there is a rapid oscillation across many (or all) of these modes of thinking. That is why it does not make much sense to characterise creativity as just *one* thing or the other, intuitive or rational, dissecting or additive when it is a fusion of these things.

Checklist

At this point we could stop here and start sorting out and evaluating our ideas for the windscreen wiper. But before doing that I would like to take you through another useful creative technique. This method involves a check list of *trigger words* - to open up new insights into a problem or different uses for ideas you've already produced.

The Checklist is given below:

Combine: Combine units? Combine purposes?
Combine ideas?

Reverse: Turn it backward? Upside down?
Transpose positive and negative?
Stop the moving? Move the stationary?

Rearrange: Interchange components? Other layout?
Other sequence? Transpose cause and effect?
Change pace?

Substitute: Other material? Other process?
Other ingredient?

Magnify: Greater frequency? Stronger?
Higher? Longer? Thicker?
Multiply? Exaggerate?

Minify: What to subtract? Smaller?
Condensed? Lower?
Shorter? Lighter?
Omit? Split up?

Modify: Change motion? Change colour?
Change sound? Change form?
Change shape? Change smell?

Put to other uses:
New ways to use as it is?
Other uses if modified?

Checklist of 'trigger' words for stimulating ideas

(Based on Osbourne, A.F., *Applied Imagination*, Charles Scribners: New York, 3rd edition, 1963, pp 286-7.)

ACTIVITY 16c
Checklist (Windscreen wiper)

Let us take some examples from the windscreen wiper problem. Perhaps the most powerful of the trigger words is *combine* because much creative thinking comes from combining previously separate objects or ideas into something new. It is the basis of the New Combinations technique.

We could use the New Combinations technique formally as described earlier in the Study Text. But instead, just think of ideas suggested by combining different parts of the windscreen and wiper system. Have a go at using the trigger word 'combine' on the windscreen wiper problem and note any ideas in your Work File.

Another useful trigger word is *reverse*.

Can you solve the problem by turning it upside down, or inside out?

Or, by stopping a moving component or moving a fixed component?

Maybe the screen could move over a fixed blade?

Or, what about a wiper on the inside of the screen to remove condensation before the heater warmed up?

I'm sure you've got the idea of the Checklist.

So now try to use the various trigger words on the windscreen wiper problem putting further ideas in your Work File.

It does not matter if you end up with good ideas for tackling a related problem - like condensation inside cars - rather than the original one.

Having produced a long list of ideas for tackling the windscreen wiper problem, the next step is to sort the ideas out and to start developing the more promising ones.

I suggest that first you try to group the ideas into broad categories as described earlier in the Study Text in the Road Safety example. I find it easier if each idea is noted on a separate piece of paper or card. Doing this classification will probably generate more ideas, so note these down as you go along.

When you have done the classification you will have a better feel for the ideas that look as if they could lead to a practical solution. But do not neglect the more way-out ideas. They can still provide a stimulus when you are working on the problem in depth.

So have a go at classifying the ideas you have produced with a view to identifying two or three promising concepts for further examination.

DISCUSSION OF ACTIVITY 16c
Checklist (Windscreen wiper)

The most interesting solutions to emerge from my classification of windscreen wiper ideas were to do with repelling the drops before they could obscure the vision by vibrating or coating the screen. Each of these ideas could be subjected to a further Brainstorming session, this time perhaps by a group with relevant expertise, before attempting a more thorough feasibility study.

However, the most potentially useful ideas were those on a related problem - that is the problem of car windows that mist up or ice over in damp or cold weather. Something that can be quite dangerous until the vehicles de-misting and heating system warms up.

It is a problem that motor manufacturers have not really solved. So it would make a good topic for a Group Brainstorming as practice for Work Task 5.

ACTIVITY 16d
Group Brainstorming (Windscreen misting)

The purpose of this Activity is to practice the procedure involved in Group Brainstorming before you use it on a real problem of your own.

To do this I suggest the following:

- Formulate a suitable question for a Group Brainstorming session on the car windscreen misting and icing problem. To remind you, the question should express the general problem to be solved in a concise but not too restrictive way.

My own suggested question:

'How can the misting or icing-up of vehicle windscreens in wet or cold weather be avoided?'

you may feel can be improved upon.

- Obtain the co-operation of at least three, preferably five to seven colleagues, plus a group leader - your manager or tutor perhaps.

- Attempt a group Brainstorming on this topic, along the lines suggested in Work Task 5. If possible make use of other creative techniques that you've learned, such as Analogies, during or after the Brainstorming.

Remember that the rules are to avoid criticism at first and to produce as many ideas as possible. Later the group can work on developing the more promising ideas into proposals for feasibility studies.

Finally, there's no need to use these techniques rigidly. Adapt them to your own way of thinking and use the ones that are most helpful to you.

IMPORTANT POINT I suggest you examine what you have learned by reviewing what you have done in your Work File, before attempting a Group Brainstorming 'for real' as suggested in Work Task 5.

WORK TASK 5

Brainstorming session

Arrange with your manager and / or tutor for about half-a-dozen people to join together in a group. This is not a technique which works well with only one or two people. Aim for about four to eight and choose *some* with knowledge of the problem area. Some companies have found it fruitful to invite participants from various departments in addition to design people. Everyone participates as equals. Those without knowledge can offer unpredictable ideas which in turn can kindle interest in the more knowledgeable. A session leader is needed who can generate a relaxed, creative atmosphere - with individuals *wanting* to offer ideas.

It is first necessary to formulate a statement of the design problem, basing it upon the information obtained in Work Task 1. Avoid vague or too general, or too restrictive statements, and do not imply solutions in the statement.

Follow this procedure for the session:

1. Allow 5-10 minutes for members of the group to write down, in silence, their first ideas in response to the problem statement. Use small cards with one idea expressed on each card.

2. Encourage group members to continue writing down new ideas, whilst *each person in turn* reads out *one* idea from their own set.

3. The rules of the session are:
   • no criticism of any idea is allowed
      (not even sniffs or raised eyebrows!)
   • crazy ideas are quite welcome
   • a large quantity of ideas is wanted
      (never mind the quality, feel the width!)
   • try to combine and improve upon the ideas of others
   • evaluation of the ideas will come after the session.

4. Close the session when ideas start to slow down - around 30 minutes should be about right.

Finally, take all the ideas and sort through them. Sort them into related groups, then look through each group for feasible ideas. Choose a few of the best ideas and try to work them up into proposals for improvements in the problem with which you started.

ACTIVITY 17 - Post-rationalising creative thinking

Having now sampled many of the techniques and hopefully begun to string them together, it is a good idea to go back and attempt one or more of the Activities again, with increased confidence.

Keep a record of your ideas in your Work File and give them some sense of order as you work. For comparison my drawings and notes on a reworking of the Beetle problem (Activity 15) are given in a summary after my Discussion of Activity 17.

DISCUSSION OF ACTIVITY 17
Post-rationalising (Beetle problem)

My thoughts were led by a series of sketches upon which I commented about possible difficulties and which led me to other ideas.

Elements
The first thought was to chase the beetle with the cylinder, so I drew that. Looking at the drawing in a more analytical mood, I realised the problem broke down into several separate elements:

- stimulus (cylinder)
- subject (beetle)
- moving and
- fixed

This gave me four elements to combine and permute.

Combination
The next thought was to attach the cylinder to the beetle, that is stimulus and subject moving together. This meant some kind of screen upon which the stimulus could be projected.

Analogue
If the beetle remained on the spot but could move (an irreconcilable polarity) then the cylinder could be fixed - stimulus and subject not moving. In my mind's eye was an image of the floating compass, which is restrained but has freedom of movement.

I thought of a golf ball suspended in liquid. This seems the right track!

Brainstorming
I was beginning to feel a little exasperated so I took it out on the beetle. What if I nailed it to the floor - or restrained it more humanely, tethered or reined back?

90

Boundaryshift
I relented to allow the beetle complete freedom of movement within a very large corral. Subject - moving, stimulus - fixed.

The stimulus would be everywhere in the beetle environment, an insect discotheque.

Boundaryshift from 4
What if the beetle was stuck on its back and its reactions measured by the waving of its legs.

This seems very obvious! I wonder why I didn't think of it before!

Boundaryshift from 2
Finally, I thought the cylinder could disappear. No drum, no environment.

Projected lights would be played directly upon the beetle in its wanderings.

This Activity took me about 15 minutes in total by which time I had run out of steam. However, looking through the sketches and notes, what became more interesting than the solutions (so called) was the actual process itself.

It was clear that one idea did lead to another in a narrative of connections; also surprisingly I had unselfconsciously used many of the creativity techniques.

The labels (Elements, Combination, Analogue, etc) were placed against the notes retrospectively.

> *There was a strong reinforcement of certain ideas which we have already heard:*

- work in a mood of relaxed playfulness
- do not be too critical or censorious
- reflect upon what you have done from time to time
- let the narrative of ideas develop
- analyse the main elements and permute them
- draw upon related objects and designs

No doubt your techniques and your narrative were in some ways different, but hopefully some of these points came across to you.

SUMMARY
Post-rationalising solutions to the Beetle problem

1 Elements
- stimulus (cylinder)
- subject (beetle)
- moving
- fixed

2 Combination
- environment moving
- stimulus projected onto screen
- mobile screen attached to beetle

3 Analogue
- environment fixed
- globe suspended in fluid
- frictionless
- golf ball
- floating compass

4 Brainstorming
- fix beetle
- nailed
- kept by bait e.g. sugar lump
- tethered on reins
- fenced in

5 Boundaryshift
- stimulus everywhere
- very large environment
- revolving lights
- discotheque

6 Boundaryshift from 4
- fix beetle on its back
- observe leg movements

7 Boundaryshift from 2
- no environment
- lights on beetle
- travelling spot

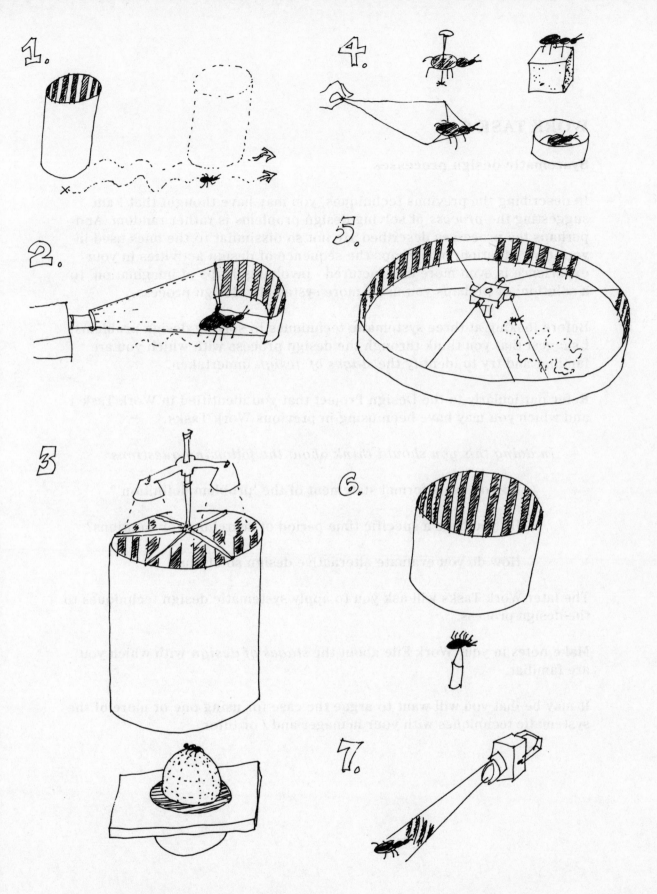

The Beetle problem. My ideas:

WORK TASK 6

Systematic design processes

In describing the previous techniques, you may have thought that I am suggesting the *process* of solving design problems is rather random. And perhaps the processes described are not so dissimilar to the ones used in your own situation; or perhaps the sequence of design activities in your experience is even more unstructured - involving 'leaps of imagination' to a solution; or perhaps you use a more systematic design process.

Before looking at three systematic techniques to solving design problems, I suggest that you think through the design process with which you are familiar and try to identify the *stages of design* undertaken.

Refer particularly to the Design Project that you identified in Work Task 1 and which you may have been using in previous Work Tasks.

In doing this you should think about the following questions:

Do you use a formal statement of the 'problem definition'?

Do you have a specific time period of searching for solutions?

How do you evaluate alternative design solutions?

The later Work Tasks will ask you to apply systematic design techniques to the design process.

Make notes in your Work File about the *stages of design* with which you are familiar.

It may be that you will want to argue the case for using one or more of the systematic techniques with your manager and / or tutor.

Objectives Tree

So far, I have described techniques involving a rather intuitive approach to solving design problems, but now I am going to concentrate on more 'systematic' approaches. My main concern is to help you apply your abilities in the best possible ways - and I hope we agree that this means being able to be 'intuitive' or 'systematic' as the situation demands. In any design process both kinds of ability need to be exercised, sometimes singly, but often by oscillating rapidly from one to the other.

The main purpose of the Objectives Tree technique is to clarify customer and other requirements when undertaking a design project. It can also be useful for generating alternative *means* of fulfilling those requirements. This technique is best applied at the early stages of a design project, when the nature of the problem is being explored; and while the design team, together with management and marketing, are trying to sort out the requirements and priorities of customers. It is especially useful for helping to reconcile market requirements with the company's design and manufacturing requirements at an early stage.

For example, customers may require a product which is reliable and safe. The design team has to decide how these objectives are to be achieved without causing conflict with the company's objective of using standard components in the design of the product. In the event of conflicts between these objectives, a compromise may have to be sought. Achieving the 'best' compromise is fundamental to producing products which can be made as simply as possible, and yet meet customer requirements (in terms of cost, safety, reliability, etc.).

The design 'brief' is the starting point for any new product. But, by its very nature, it is usually a very brief statement of requirements. The writer of the brief (usually a marketing person in the case of a manufacturer of consumer products) may not be able to 'specify' the objectives in the terms which will be helpful to the designer. To understand this better, let us look at an example of how objectives can be better clarified.

In the case of a lawn-mower, the design brief may include key objectives such as, 'the product must be safe and reliable'. This kind of statement gives you little help in establishing the boundaries or parameters within which you can work. It does not, for example, tell you 'how' safe or 'how' reliable the lawn mower should be.

So how can we expand and clarify these rather simplistic objectives? Using adverbs, such as, Why? What? Where? How?, can be very valuable in developing questions which elicit other objectives.

For example, for a lawn-mower, 'What is meant by reliable?'.

> *As a result of asking this question, the objective of 'the product must be reliable' can be developed into:*
> - High number of operations (i.e. times of use) without attention
> - High cutting use without attention
> - Low wear of moving parts
> - Low level of vibration
> - Low risk of misuse

The process of developing a list of objectives will usually result in having different levels of objectives. For instance the 'means' of achieving an objective will be an objective in itself - but a lower level of objective. For the example of the electric lawn-mower, the means of achieving the objective 'Works first time for the first 1000 operations' may require the lower level objective, 'Automatic overload device' in order to protect the electric motor from damage.

The process of developing objectives in this way will result in a *branching* of objectives from the original key objectives. In this way, several levels of objectives will result. At each level the objectives will be the 'means' of achieving the objective at the level above. Considering our lawn-mower again, a list - with groups of objectives at each level - will result:

- Product must be reliable (Key objective)

- High number of operations without attention
- High cutting use without attention

- Low wear of moving parts
- Low risk of vibration
- Low risk of misuse

As you write down and expand the lists under each key objective, it may be that some of the 'means' objectives will be able to meet more than one of the higher level objectives. For example, the 'means' objective 'Low risk of misuse' may contribute to achieving the product's requirement not only for reliability but also for safety. In order to see more clearly the links and dependencies between each of the objectives, a 'tree' diagram can be drawn. (Hence the name of the design technique - 'Objectives tree').
The tree is usually drawn with the key objectives at the top and the lower objectives forming the lower branches. In each case the links between objectives are drawn to show the lower objectives which are the means to achieving the higher objectives. The best way of showing what I mean is to once again look at the example of the lawn-mower, but this time to look at more key objectives.

96

USING THE OBJECTIVES TREE TECHNIQUE

EXAMPLE: LAWN-MOWER

You will see below an objectives tree for a lawn-mower which is required to be 'safe, reliable, convenient to use, and simple to make'. Each of these rather woolly objectives is made more explicit at each level down the tree by including the 'means' of achieving them.

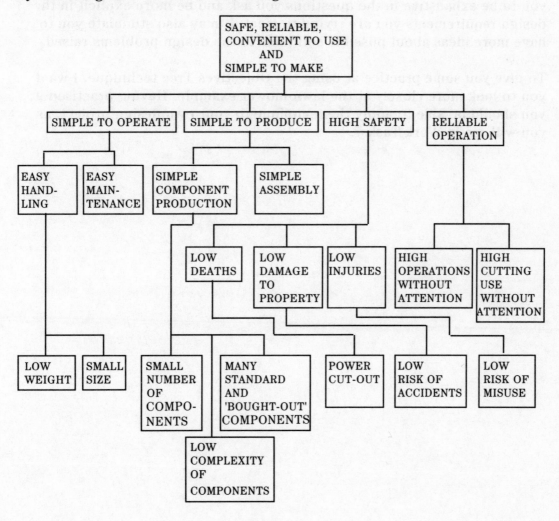

An objectives-means tree for a 'safe, reliable, convenient to use and simple to make' lawn-mower.

You may well already have your own methods of establishing the product's design requirements. The Objectives Tree technique is not intended to replace these methods, but to provide a systematic approach to the setting and meeting of design requirements. In using this approach, it may be that different people will think of different 'means' and different links between the objectives. However, the process of developing the tree should cause you to be exhaustive in the questions you ask and be more explicit in the design requirements you are trying to meet. It may also stimulate you to have more ideas about possible solutions to the design problems raised.

To give you some practice at using the Objectives Tree technique, I want you to look more closely at the lawn-mower example. Having practised it, you should be able to apply the technique to your own design project, as you will see in Work Task 7.

ACTIVITY 18a
Objectives Tree

Below is part of the Objectives Tree for the lawn-mower example. Look at the objective 'Reliable operation' on the right-hand side of the tree, and the possible *means* of achieving this key objective. It is the part of the tree enclosed by the dotted line.

Each box either represents an objective to be achieved by different 'means' at the next level down, or is the 'means' of achieving the objective at the next level up. For example, 'High cutting use without attention' is both an objective and also a means of achieving the objective 'Reliable operation' at the next level up.

Assuming the Design Brief for the lawn-mower states:

> 'A design is required for a 1 kW lawn-mower, powered by mains electricity, not being too heavy to carry, able to collect grass cuttings and quiet in operation. Attractive to people with small lawns to cut, easy to handle and operate, and convenient to store in the minimum amount of space'.

Consider the different means of achieving the objective 'high number of operations without attention'. Using your Work File enter your ideas by extending the Objectives Tree shown enclosed by the dotted lines. When you have done this, look at my suggestions in the Discussion of Activity 18a on the following page.

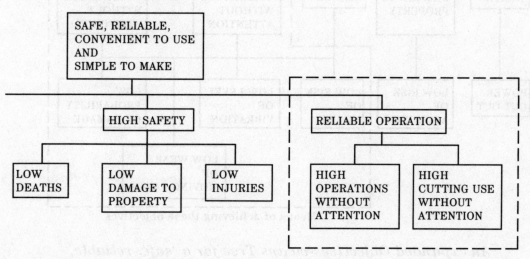

Lawn-mower example - part of the Objectives Tree

DISCUSSION OF ACTIVITY 18a
Objectives Tree

You may well have had different ideas from mine shown below. This does not really matter! What matters is the process of systematically considering each objective and the means by which it can be achieved. In a real design situation this process would involve consultation with customers, retailers, marketing people and other departments in the company, as well as other design team members.

The method forces everyone involved to be more explicit about their objectives and the different ways of achieving them. Having done this it is possible to further clarify the objectives in order to transform them into performance criteria. The sort of criteria which will enable you to actually start to specify - in much more concrete terms - the parameters within which you can make actual design decisions.

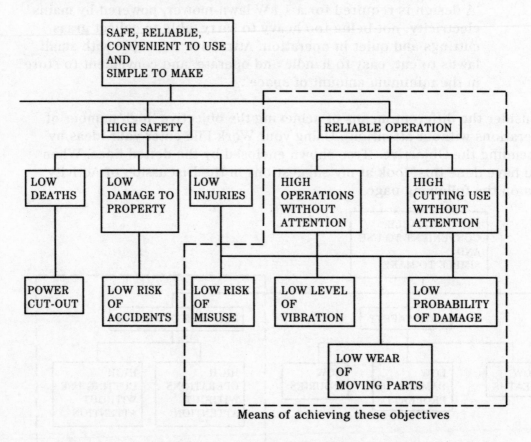

Means of achieving these objectives

An expanded Objectives-means Tree for a 'safe, reliable, convenient to use and simple to make' Lawn-mower (my version)

ACTIVITY 18b
Translating objectives into a Performance Specification

Choose three or four objectives at the third level down (or lower) from the Objectives Tree for the lawn-mower - for instance, 'Low wear of moving parts'. Then try to define these more precisely as measurable criteria, against which you can assess an acceptable design of a lawn-mower having low wear of its moving parts.

Enter your thoughts in your Work File.

When you have done this, read the Discussion of Activity 18b on the next page. I have given my suggestions for how to express as performance criteria the objectives 'Low wear of moving parts', 'Low probability of damage', 'Low level of vibration' and 'Low weight'. If you choose any of these objectives, do not look at my suggestions before trying this activity.

DISCUSSION OF ACTIVITY 18b
Performance Specification based upon objectives

As with the original Objectives Tree there are no right answers to this activity, so my suggestions given below are likely to be different from yours. Defining the performance criteria has to be done in consultation with all those involved in your design project and may require a fair amount of investigation before precise data can be included.

I will not discuss this any further now as the development of a *performance specification* is discussed in detail after you complete Work Task 7.

Performance Specification based upon Objectives

Objective	Measurable performance criteria
1. Low wear of moving parts	Switch will operate 10,000 times without fault. Electric motor will operate for 250 hours without fault. Cutting blade will cut effectively for 1,000 hours without attention.
2. Low probability of damage	Body of lawn-mower can withstand a side impact without cracking as a result of being dropped from a height of 1 metre under the influence of gravity.
3. Low level of vibration	Static out-of-balance of cutting blades not to exceed 2 grams.
4. Low weight	Total weight of machine not to exceed 50 kg.

WORK TASK 7

Clarifying customer requirements

Ideally, in order to undertake this Work Task, you will need to be in a situation where a change in product design is necessary to meet new or changed market requirements. You will need to discuss with your manager and/or tutor the requirements for the task outlined below.

Note If you are not at present in the situation described, then there is much benefit to be gained from undertaking this task on an existing product - by imagining that the product has to be redesigned from scratch.

1. Obtain a 'brief' statement of design objectives for the product from an appropriate person - this may be your manager, tutor, marketing manager, sales manager, customer, or whoever.

2. Clarify, expand and revise this 'brief' statement into a more complete list of objectives, by asking yourself (and others) appropriate questions.

3. Sort the list into sets of higher-level and lower-level objectives and the 'means' of achieving these objectives.

4. Draw an objectives tree in diagrammatic form showing the hierarchical relationships and interconnections between the different levels of objectives and means.

The results of undertaking this task should be entered in your Work File for later discussion with your manager and / or tutor.

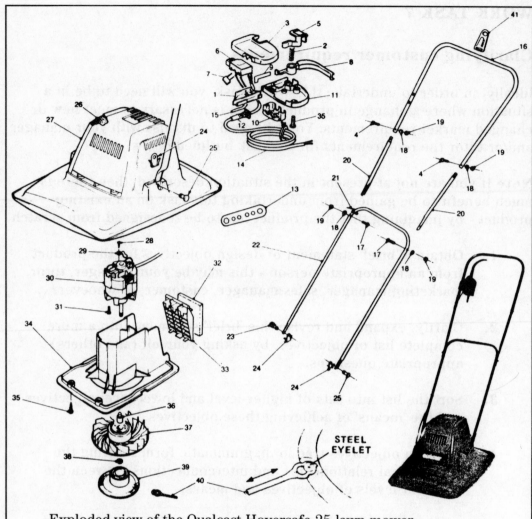

Exploded view of the Qualcast Hoversafe 25 lawn-mower

Parts list by item number and description:

1 Handle clamp	15 Intermediate cable assembly	29 Cable clamp
2 Spring (buttton return)	16 Handle top	30 Motor assembly
3 Switch housing upper	17 Bolt handle securing (2off)	31 Screw (motor mounting) (4off)
4 Bolt (handle securing,2off)	18 Shakeproof washer (4off)	32 Filter
5 Button (safety)	19 Nut (handle securing) (4off)	33 Filter retainer
6 Switch	20 Handle centre (2off)	34 Motor mounting plate
7 Capacitor	21 Cable cleat	35 Screw (12off)
8 Switch lever	22 Handle bottom	36 Adaptor
9 Spring (lever return)	23 Looped pivot pin	37 Fan
10 Cable clamp	24 Hinge butt (2off)	38 Ratchet plate
11 Switch housing lower	25 Pivot pin	39 Cutter disc/locking wheel assy.
12 Cable guard	26 Foam seal	40 Cutter
13 Spanner	27 Deck	41 Cable restraint
14 Supply cable assembly	28 Screw	

An electrical mains-powered rotary hover lawn-mower with a plastic cutting blade

(*Courtesy:* Qualcast Garden Products Ltd.)

Performance Specification

The process of planning for a new product ultimately leads to the need for a 'product definition'. In other words, the 'specification' of the most important requirements and features of the final product. The product definition is usually expressed in the form of a performance *specification* or *requirements* list.

For a designer, the *product definition* represents the initial definition of the problem to be solved. The 'product definition' is therefore the 'problem definition' to which you have to find design solutions. More particularly, you are required to attempt to find the optimum solution within the constraints given. As we discussed when looking at the 'Boundary Shifting' technique, establishing the boundary or constraints of the design problem is often not easy and may be subject to change as the product definition is refined.

A simple specification list or performance requirement list is a necessity when having to satisfy particular criteria for a new product. Such lists represent your ultimate source of reference - even though the product definition may change - when examining the acceptability of proposed design solutions. Having a product definition based upon the best available information and comprehensively covering all the important requirements provides you with the confidence of knowing the right problem is being tackled. To ensure that the information is correct and complete requires you to have close contact with your client or customer. The problems of establishing good communication with your client / customer and all the others concerned in defining the product is not the subject of this book.

But, suffice it to say that it is crucial to get the right answers to questions, such as:

- How was the design project originated?

- What is the problem?

- What are the wishes and expectations of those involved?
  (e.g. user, customers, marketers, production engineers)

- Do the given constraints actually exist?

- What performance attributes and requirements **must** the product have and not have? Which are desirable but not essential?

When obtaining the answers to these questions it is important to establish the *level or levels of generality* at which you are required , or able, to work. This is the first stage of using the Performance Specification method. For a new product this means:

- Specifying the product alternatives (high level of generality)

- Specifying the product types

- Specifying the product features (low level of generality)

Let's look at some examples of specifying a new domestic cooking appliance:

Alternatives
Imagine the alternative ways of cooking food e.g. baking, frying, stewing, boiling, steaming, grilling, micro-waving. Or, perhaps an entirely new way of cooking. This is the highest level of generality and will be appropriate if your company wishes to diversify or change the basic concept of the appliance it manufactures.

Types
At this level, we will need to consider the basic differences between the various types of cooking appliance - oven, grill, hot plate; and / or the types of fuel to provide the heat - solid fuel, oil, gas, electricity.

Features
This is the lowest and most detailed level. In this, we consider the features of the particular type of appliance e.g. size of heating element, cooking capacity, temperature control, etc.

The level of generality at which you can work will be determined by your company's senior management or perhaps your customer. Obviously, the higher the level of generality, then the greater the freedom you will have to make major design decisions. Having established the level at which you are able to work, you can start preparing a performance specification. This means listing the performance attributes - rather like listing the design objectives in the Objectives Tree method.

IMPORTANT POINT The performance attributes and requirements should not contain statements that prescribe the design solution - whether wholly or in part. The list of statements should only involve the conditions that the design solution must satisfy.

Let's look at an actual example:

USING THE PERFORMANCE SPECIFICATION TECHNIQUE

EXAMPLE: ONE-HANDED MIXING TAP

This problem exists at the lowest level of generality - so it is concerned with the *features* of a one-handed mixing tap. The first task was to list the performance attributes to be met:

- various water flow rates

- various water pressures

- various water temperatures

- various connections

- various distances (outflow to basin edge)

- various loads to operate tap

- various fitting requirements

- various water supplies

- independent of external heating

A general description of the problem is then formulated into a specification statement:

'Regulate the temperature of water supplied to a wash hand-basin from water drawn from household hot and cold main supplies and controlled by a mixing tap requiring one hand and light in operation'.

This statement is in turn translated into a detailed performance specification:

- Maximum mixed flow rate 10 litres / minute at 2 bar pressure

- Maximum water pressure 10 bar

- Water temperature 60 °C (standard),
  capable of 100 °C (short time)

- Temperature setting independent of flow rate and pressure

- Temperature fluctuation permitted of + / - 5 °C at a pressure of
  + / - 5 bar between hot and cold supply

- Connections: 2 off copper pipes, 10 mm dia., 400 mm in length

- Outflow of tap: 50 mm above upper edge of basin

- Tap to fit household basin

- Light operation for use by children

- Handle not to heat above 35 °C

In working on design problems you may find it useful to use a standard form when finalising the performance specification. This may then also subsequently be updated as any changes to the specification are made. The specification form on the adjacent page shows the actual results of a team of designers in developing the requirements for the one-handed household water mixing tap. You will notice that, in this case, it is not just a performance specification, but includes other requirements. For example, the requirements for the trade mark (Item 15), the cost and rate of manufacture (Item 26) and the project timetable (Item 27).

Performance Specification-
One-handed household water mixing tap

Changes	D/W	Requirements				SPECIFICATION for One-handed mixing tap — Page 1	Resp.
	D	1 Throughput (mixed flow) max 10 l/min at 2 bar					
	D	2 Max. pressure 10 bar (test pressure 15 bar as per DIN 2401)					
	D	3 Temp. of water: standard 60°C, 100°C (short-time)					
	D	4 Temperature setting independent of throughput and pressure					
	W	5 Permissible temp fluctuation ±5°C at a pressure diff. of ±5 bar between hot and cold supply					
	D	6 Connection: 2 × Cu pipes, 10 × 1 mm, l = 400 mm					
	D	7 Single-hole attachment Ø 35±? mm, basin thickness 0–18 mm (Observe basin dimensions DIN EN 31, DIN EN 32, DIN 1368)					
	D	8 Outflow above upper edge of basin: 50 mm					
	D	9 To fit household basin					
	W	10 Convertible into wall fitting					
	D	11 Light operation (children)					
	D	12 No external energy					
	D	13 Hard water supply (drinking water)					
	D	14 Clear identification of temperature setting					
	D	15 Trade mark prominently displayed					
	D	16 No connection of the two supplies when valve shut					
	W	17 No connection when water drawn off					
	D	18 Handle not to heat above 35°C					
	W	19 No burns from touching the fittings					
	W	20 Provide scalding protection if extra costs small					
	D	21 Obvious operation, simple and convenient handling					
	D	22 Smooth, easily cleaned contours, no sharp edges					
	D	23 Noiseless operation (≤20 dB as per DIN 52218)					
	W	24 Service life 10 years at about 300 000 operations					
	D	25 Easy maintenance and simple repairs. Use standard spare parts					
	D	26 Max. manuf. costs DM 30 (3000 units per month)					
	D	27 Schedules from inception of development					
			conceptual design	embodiment design	detail design	prototype	
		after	2	4	6	9 months	

The D/W column shows whether the specification requirement is a
'Demand' or a 'Wish' - that is, it distinguishes between 'essential'
requirements and 'desirable' requirements.

(*Source*: G. Pahl and W. Beitz, *Engineering Design*, The Design Council, London, 1984.)

When listing the performance attributes it is very easy to end up with a 'wish list', a list of all the things you would like to satisfy in the design solution. However, this can present unnecessary problems when trying to satisfy a customer or user whose basic demands are for a product which has certain attributes which are crucial and at a price which is affordable.

To help in the process of establishing the 'hard' and 'soft' attributes, it may prove necessary to establish which are 'demands' and which are 'wishes'. For example, the requirement for a kettle to boil a litre of water may be a demand, but the requirement to measure the amount of water contained at various levels may be a 'wish' to be satisfied only if it can be achieved within a certain cost limit. If you refer back to the Performance Specification for the *one-handed mixing tap*, then you will see each item identified as either a 'demand' or a 'wish'.

A performance specification attempts to define fairly precisely what a satisfactory design must do, but not what form it may take. Having listed each attribute, it should be possible to write a *specification statement* which clearly describes the problem. This statement should then be broken down into the performance specification - which will itemise in quantifiable terms (wherever possible) the performance criteria to be met.

For example, do not be satisfied with simply stating that a lawnmower 'must be portable', but say what this means in terms of its maximum size and weight. It is also important to establish the limits within which an acceptable performance lies. So do not state, "The Cutting Speed must be 4,500 metres / minute' if a cutting speed of anything between 4,000 and 5,000 metres / minute is acceptable. However, it is important to allow the maximum freedom for your design. So do not give an exact dimensional specification if it is not necessary. For example, do not specify, 'The housing must have a surface texture of 10 to 50 μm.*', when what you mean is that, 'The housing must be smooth to the touch'.

(* The surface texture or surface finish is a measure of 'roughness'. In the case of injection moulded plastics it indicates the acceptable limits of roughness of the steel mould used and is measured in micrometres - μm.)

ACTIVITY 19
Performance Specification
(Human-powered transport)

A Performance Specification attempts to define fairly precisely what a satisfactory design must do, but not what form the design will take. So first, I would like you to define what *performance attributes* or *requirements* you would expect a human-powered transport device to have. They are rather like the objectives in the Objectives Tree technique.

Write down in your Work File three or four essential attributes that would be required in an acceptable design. Don't look at my Discussion of Activity 19 on the next page until you have attempted this for yourself.

When you have compiled a realistic list of attributes you can begin to write performance criteria based on each. These define more precisely the limits of acceptable and unacceptable designs.

Defining the performance requirements in a precise enough form to provide a useful specification may also require careful investigation.

Suppose now we want to define more precisely a particular attribute for a human-powered transport device, such as, 'light enough to carry with ease'. What questions will you need to investigate to define this attribute more precisely? Write down your questions in your Work File.

Having done this, read the Discussion of Activity 19 on the next page and compare your questions with mine.

DISCUSSION OF ACTIVITY 19
Performance Specification
(Human-powered transport)

ESSENTIAL PERFORMANCE ATTRIBUTES FOR A PORTABLE, HUMAN-POWERED TRANSPORT DEVICE FOR USE IN CONJUNCTION WITH A CAR OR TRAIN *My suggestions:*

1. Light enough to carry with ease

2. Compact enough to fit into a car boot or inside a train compartment

3. If it is to fold, this should be quick and easy

4. Comfort and speed not greatly inferior to a conventional bicycle (at least for short trips)

The performance attributes that I have listed above are ones which came fairly readily to mind when thinking about a human-powered vehicle. But, if you were being rigorous about compiling such a list, you would probably have to investigate more fully the requirements of potential customers and the market. To do this, critical questions may need to be asked about each attribute. In order to give you practice in defining more precisely the limits of acceptability of designs, you were asked to look at the attribute, 'light enough to carry with ease' for the human-powered transport device. Listed below are my suggestions for the critical questions to be asked.

QUESTIONS TO BE INVESTIGATED IN ORDER TO DEFINE MORE PRECISELY THE ATTRIBUTE 'LIGHT ENOUGH TO CARRY WITH EASE' *My questions:*

1. Who is to do the carrying (men, women, children) ?

2. How far is the device to be carried ?

3. What is the maximum weight users would find acceptable to carry for the expected distances ?

DISCUSSION OF ACTIVITY 19 (Continued)
Performance Specification

The specification that may emerge from investigating the requirement 'light enough to carry with ease' may be, 'light enough for an adult male to carry for a distance of up to 200 metres without discomfort'. Looking this up in a hand book of human factors data may give a maximum weight for this requirement (i.e. the vehicle) of about ten kilograms.

This is still not as rigorous a specification as some situations demand. Safety requirements for instance must be very precisely defined and may be laid down in *standards* or *codes of practice* - for example, the British Standard giving the maximum stopping distance for cycle brakes under given conditions.

IMPORTANT POINT The performance specification does not say, how the requirements - for example, the vehicle's light weight - are to be achieved or even what sort of device it is!

This leaves you free to decide on different solutions that meet the defined performance specification.

Inevitably there will only be a few solutions that meet the whole specification and in this way the options may be narrowed down.

In fact the performance specification I have begun to develop is similar to that produced by the designer of the world's first fully portable bicycle. Harry Bickerton used his performance specification - it is reproduced on the next page - as the basis for design and prototype studies of novel forms of human-powered vehicles.

In the end he found the design that best met the 'specification' was a small-wheel collapsible bicycle made from light alloy materials. There is an illustration of it on the next page.

NOTE Before attempting Work Task 8, I suggest that you review what you have studied about Performance Specifications by looking back through your Work File.

PERFORMANCE SPECIFICATION FOR A NEW PORTABLE HUMAN-POWERED PERSONAL TRANSPORT DEVICE

1. The vehicle must be as light and as compact as possible and in any case no more difficult for a pedestrian to carry or stow away than a medium-sized suitcase.

2. It must enable its user to cover distances up to, say 5 miles more easily and at the very least four times faster than fast walking on prepared surfaces such as roads.

3. It must not require standards of physical ability and skill above normal.

4. It must be mechanically safe, reliable, and roadworthy under all normal conditions.

5. Time and effort required to change from the 'carrying' to the 'riding' configuration and vice-versa to be a minimum and time, in any case, not to exceed one minute.

6. It must be suitable for all sizes of people covering the age of 10 years and above, and for both sexes.

7. The appearance must be attractive and remain so in use and storage over, say, five years.

8. It must be suitable for quantity production and sell at an economic price which should not in any case be more than twice that of a good quality bicycle.

9. It must require the minimum of maintenance and no special care or attention.

10. It must be suitable for worldwide distribution, sales and servicing.

(*Source:* H.Bickerton, *Bickerton portable bicycle*, Advance Information, December1972)

The design which best met the performance specification was the Bickerton - the world's first fully-portable bicycle. It weighed only 18 lb (8.2 kg) for the single-speed version due to extensive use of light alloy materials. The Bickerton folds into a space of 685mm X 533mm X 254mm.

WORK TASK 8

Specifying performance requirements

As with Work Task 7 involving the Objectives Tree technique, ideally you should be in a situation where the design of a new product is required. But the Performance Specification technique can also be applied to an existing product, if you are seeking to evaluate the existing design or to generate some redesign proposals.

If you carried out Work Task 7 (Objectives Tree), then you should find that a useful starting point for this Work Task.

1. In discussion with your manager and/or tutor, decide on the *level of generality* at which you will operate - i.e. whether you are concerned with 'alternatives', 'types' or 'features'.

2. Using your previously-generated list of objectives, or other information you have obtained on the requirements for your product, draw up a list of the product's *performance attributes*.

3. Choose one or two of the performance attributes and carry out the necessary research or testing to enable you to write precise *performance specifications* for those attributes.

4. Write-out as much of a complete performance specification as you can for your product, using a format like the examples shown earlier.

The results of undertaking this task should be entered in your Work File for later discussion with your manager and / or tutor.

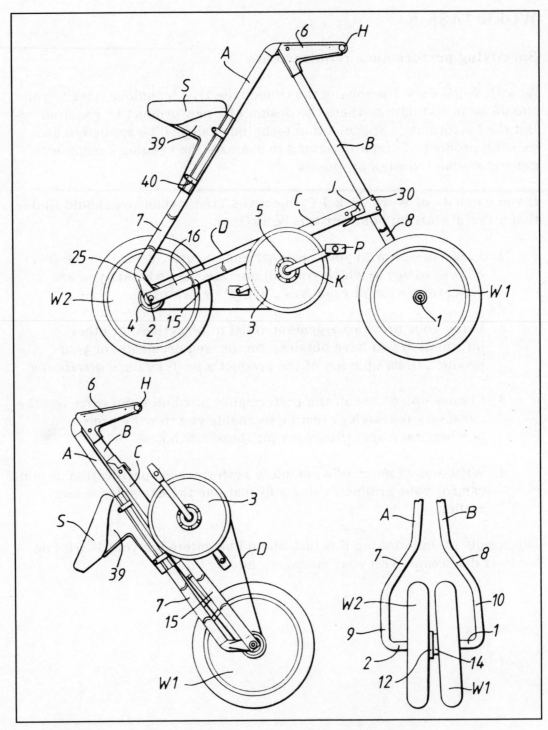

Strida collapsible bicycle invented by Mark A. Sanders
(Source: U.K. Patent Application GB 2171656A, September 1986)

The drawings reproduced here are taken from the filed patent application.

Morphological Analysis

The Morphological Analysis technique is particularly useful when seeking to identify new concepts involving the configuration of the total product. The word 'Morphology' means the study of *form* - by which we mean general configuration. For many design problems, the search for design solutions involves a consideration of possible 'forms' or 'configurations'. The technique provides a systematic approach to combining basic design concepts in ways which may not have otherwise been considered.

To achieve this, there are two main tasks:

- Systematically establish the *basic parameters* of the design problem.

- Generate *individual solutions* for each parameter by applying the principle of association or variation.

In an attempt to ensure that all possible solutions are considered, the technique involves *charting* all the possible forms in a methodical way. This means using a matrix to construct a 'morphological chart'. Possible solutions are entered on the matrix as combinations of sub-solutions. You are then able to select alternative sets of sub-solutions, ending in the probability of arriving at new combinations which would not have otherwise been considered. This is a particularly useful technique when searching for design solutions to a new product concept, reviewing the existing product concept or examining competitors' products.

Let's look closely at what this means step by step:

- List all the essential parameters, that is, the *functions*, *features* and *characteristics* that you wish the product, process or service to possess.

- You should consider the parameters as being sub-problems of the total problem to be solved.

Note The list should be six to eight items in length, each having the same level of generality and being as unrelated to each other as possible.

- Search for alternative solutions to each parameter by using the principle of association or variation. (Refer to the 'New Combinations' technique to remind yourself of this principle)

- List your solutions. Each solution will represent the *means* by which each function, feature or characteristic can be achieved through the product's design. Or, put another way, each means will be a sub-solution to a sub-problem of the total problem to be solved.

Note These lists should identify the physical medium or constituent component necessary to achieve the function, feature or characteristic. To understand this better, look at the example 'Concept Scheme for Refuse Separation Equipment' on the next page. The 'means' may include not only the existing well-tried ones, but also entirely new ones - so long as they are realistic possibilities.

- Construct a matrix of empty boxes. The matrix will need to be as high as the number of essential 'parameters' you have listed; and wide enough to be able to enter all the possible 'sub-solutions' in achieving the parameters.

- Enter each item from your list of parameters (i.e. functions, features and characteristics) in the left hand vertical row of boxes.

- Enter each item from your lists of sub-solutions (i.e. the 'means' of achieving each function, feature and characteristic) across the appropriate row.

Note The number of 'means' of achieving a particular function, for example, may be different from another. To understand this better, if you look at the example 'Forklift truck' on page 122, you will see that there are five 'possible solutions' to the parameter 'support' and only three possible solutions to the parameter 'steering'.

The completed matrix will be a morphological analysis containing all the possible configurations for the product.

- Select one sub-solution at a time from each row and so arrive at different sets of combinations, that is, *total solutions*.

Note In this way all the possible total solutions can be identified. The number of possible combinations can amount to a very large number and will need to be reduced to a manageable number. To understand this better, look at the example, 'Concept Scheme for Refuse Separation Equipment' where three quite different combinations are identified.

- List each set of possible combinations of *sub-solutions* - provided that it is not too many.

Note If the number of combinations is too large, then eliminate all the unrealistic and/or incompatible pairs of sub-solutions as you proceed by not listing them.

- Consider each set of combinations of sub-solutions (i.e. total solution) in turn.

Note As a result of further consideration you may wish to further reduce the solution list because of economic, manufacturing or other practical reasons.

The result of all this effort should be a short-list of possible total solutions worthy of further in-depth feasibility studies. This short-list will undoubtedly have possible solutions that you would have in any case considered. But, despite this, you may end up with ideas you would not have thought of, and have greater confidence in the solutions to be studied.

Having described the process of developing a morphological chart, let us look again at two examples of this technique. If you are still uncertain about this process, you may wish to refer back to the steps I have described whilst studying the examples.

USING THE MORPHOLOGICAL ANALYSIS TECHNIQUE

EXAMPLE: CONCEPT SCHEME FOR REFUSE SEPARATION EQUIPMENT

Deciding the concept for equipment best suited to separating refuse is a problem which can be analysed using the morphological technique. The configurations of such equipment require an examination of seven basic 'parameters':

- Refuse collection
- Transportation
- Separation
- Storage container / room
- Analyse composition of refuse
- Comminution (to pulverise)
- Storage site

It was decided that the possible 'means' (in this case the physical medium) of fulfilling the function of refuse collection could be:

- Conveyor belt
- Flap
- Container (stationary)
- Hopper
- Bunker

This process of analysis was carried out for each of the parameters (i.e. sub-problems) and lists of solution options (i.e. the 'means') were drawn-up. An appropriate matrix was constructed and the resultant morphological chart is shown below:

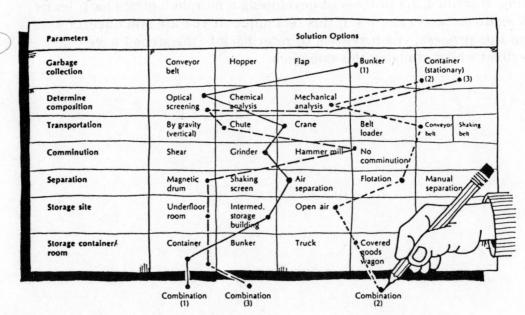

Concept Scheme for Refuse Separation Equipment
(Source: *Idea Generation Methods - Creative Solutions to Business and Technical Problems*; Geschka, von Reibnitz and Storvik 1982)

If you chose to calculate the number of different combinations possible from the chart, you would find that there are 21,600 possible configurations for the concept scheme. Of course, many of these combinations will not be realistic or feasible.

Three feasible total solutions are illustrated in the table, they are:

Combination 1 - Refuse separation equipment (concept scheme)

- Bunker for collecting the refuse
- Optical screening of the refuse
- Crane for transporting the refuse
- Grinder for comminution (pulverising) of refuse
- Air separation facilities
- Storage containers for roofed intermediate storage

Combination 2 - Refuse separation equipment (concept scheme)

- Stationary containers for refuse collection
- Mechanical analysis of refuse composition
- Conveyor belt for moving refuse
- Flotation facilities for separation
- Storage wagons for immediate outdoor storage

Combination 3 - Refuse separation equipment (concept scheme)

- Stationary containers for refuse collection
- Optical screening of the refuse
- Gravity chute for transporting the refuse
- Magnetic drum for separating iron-based refuse
- Storage containers for underfloor room storage

Perhaps you would like to try to identify two or three other total concept schemes by listing the combinations involved. This means looking for configurations which are realistic and feasible from the 21,600 possible combinations. Perhaps a daunting task but a little practice in thinking about *new combinations* in a systematic way will prove helpful in trying the technique for real.

EXAMPLE: FORKLIFT TRUCK

Designing vehicles of a configuration best suited to lifting, carrying and placing loads of various sizes and weights needs an examination of a very complex range of alternatives. Such a vehicle is a forklift truck. The following list of 'parameters' includes the essential *functions*, *features* and *characteristics* that a forklift truck should possess.

- Support - the means of supporting the vehicle in achieving its mobility
- Steering - the means of steering the vehicle
- Stopping - the means of stopping the vehicle
- Moving - the means of propelling the vehicle
- Power - the means of providing energy for propulsion
- Transmission - the means by which power is transmitted into propulsion
- Lifting - the means by which loads can be lifted
- Operator - the means by which the operator is located

As a result of considering the 'possible sub-solutions' to achieving these parameters, the following morphological chart can be drawn up:

Para-meters	Possible sub-solutions				
Support	Wheels	Air cushion	Tracks	Slides	Spheres
Steering	Turning wheels	Rails	Air thrust		
Stopping	Reverse power	Brakes	Blocks under wheels	Drag a weight on the floor	
Moving	Air thrust	Power to wheels	Hauling along a cable	Linear induction motor	
Power	Electric	Bottled gas	Petrol	Diesel	Steam
Trans-mission	Hydraulic	Gears and shafts	Belts or chains	Flexible cable	
Lifting	Screw	Hydraulic ram	Rack and pinion	Chain or rope hoist	
Operator	Seated at front	Seated at rear	Standing	Walking	Remote Control

Morphological chart for a forklift truck
(Source: Pitts, G, *Techniques in Engineering Design*, Butterworth, London,1973)

If you calculate how many possible solutions there are as a result of this analysis, you will find it totals 96,000. But many of these vehicle configurations will be impractical, or suggest incompatible solutions - e.g. a tracked vehicle can hardly be steered by wheels!

The biggest difficulty in making the morphological analysis effective is in the selection of the parameters to the problem. The chosen set of parameters represent the *basic concept* of the product, process or service. It is very important to avoid introducing parameters which are not basic and therefore do not contribute to reaching novel solutions to the problem.

As previously discussed, only six to eight parameters should be used. This may mean compiling a list of only the most important aspects of the problem. The individual parameters must be important to the basic concept (and therefore the problem) and not overlap - thereby allowing the widest variety of solution options.

The steps in undertaking the morphological analysis of a problem can usually best be done by a group of people. The stimulation of the suggestions from other group members should ensure that all aspects of the problem parameters are explored and a wide range of solutions generated.

During the initial analysis of the parameters, the use of pin cards and a pin-board can be very effective. This involves group members writing their suggestions for parameters on cards, which are then pinned to a board. All duplicate parameters are removed; and those parameters which overlap are reformulated and new cards exchanged.

Ideally, the brainstorming of a group should be used in developing solution options. Again, following the brainstorming discussion, pin cards can be used to record the various options and then pinned directly to the morphological chart. (Note: You may wish to refer to the technique of Brainstorming described earlier in the Study Text.)

The final identification of feasible total solutions (i.e. feasible combinations of sub-solutions) requires much more than just combining the alternative sub-solutions together. It requires a careful judgement based upon a wide range of factors including:

- Cost of manufacture
- Existing manufacturing resources
- Interchangeability with existing parts
- Changes to existing servicing facilities and methods
- Available investment resources
- Distributor / user acceptability

..... and many others.

IMPORTANT POINT The quality of the decisions made during this last step is critical to arriving at solutions which are not only novel but also feasible and therefore worthy of further investigation.

Modern hydrostatic drive forklift truck
(*Courtesy:* Boss Trucks Ltd.)

ACTIVITY 20
Morphological Analysis (Forklift Truck)

In order to gain a real insight into the purpose of this technique, I would like you to examine more closely the previous example of the forklift truck. Redraw the morphological chart shown on page 122 in your Work File.

Part (a) Carefully study the chart and select a conventional forklift truck vehicle configuration. Do this by entering your ideas in your Work File. Don't worry if you are unfamiliar with forklift trucks. This will involve choosing one sub-solution from each row on the chart. The resultant set of sub-solutions should be compatible with each other and when considered together, you should feel they best describe the configuration of a conventional forklift truck. Note: You have plenty of possible combinations from which to choose - there are 96,000.

Spend 5 - 10 minutes on Part (a) of this activity, making notes in your Work File. Do this before going on to read a Discussion of Activity 20 Part (a) on the next page and before doing Part (b).

Part (b) Again, study the chart and this time try to select unconventional vehicle configurations. This should involve trying to combine compatible sets of sub-solutions that describe a new vehicle configuration which is also a feasible *total design solution*. However, to make this task more manageable and to give you practice developing your own sub-solutions, I suggest you develop your own morphological chart. Base the chart upon the same parameters, but this time add some of your own sub-solutions, and then use it to select unconventional vehicle configurations. Do this for two or three sets of alternative combinations and give your reasons as to why you consider them to be feasible.

Spend 30 - 45 minutes on this activity, enter your ideas in the form of your own Morphological Chart in your Work File. Do this before going on to read a Discussion of Activity 20 Part (b) on the following page.

DISCUSSION OF ACTIVITY 20
Morphological Analysis (Forklift Truck)

Part (a) You were asked to select a conventional vehicle configuration for a forklift truck.

These are my own selections from the chart:

CONVENTIONAL FORKLIFT TRUCK	
Parameters	**Sub-solutions**
Support	wheels
Steering	turning wheels (usually rear wheels)
Stopping	brakes
Moving	power to wheels
Power	electric (or petrol / diesel fuel or bottle gas)
Transmission	gears and shafts
Lifting	rack and pinion (or hydraulic ram or chain)
Operator facilities	seated at front (or walking)

If, as a result of being unfamiliar with forklift trucks, your own version differs wildly from mine, then you may find it useful to locate a vehicle and carry out a close examination.

DISCUSSION OF ACTIVITY 20 (Continued)
Morphological Analysis (Forklift Truck)

Part (b) You were also asked to use the chart to identify less conventional vehicle configurations. Your ideas will probably not be the same as my own, but here is one of my own efforts and my reasons for selecting the sub-solutions chosen.

I began by considering the size and weight of the loads to be lifted. As a result, I concluded that very heavy loads may require a quite different vehicle configuration than lighter loads. Following this line of thought, I chose to consider the prospect of a small and highly manoeuvrable lifting vehicle capable of carrying loads which are too heavy for a person to reasonably carry. The need for such a vehicle would exist in many warehouses and large offices where loads need moving from one location to another and lifted and lowered for dispatch purposes. Conventional forklift trucks of all sizes are heavy in relation to the loads they lift in order to counterbalance the downward force of the load on the lifting forks. This led me to think of the possibility of a lifting vehicle with a configuration which would be lighter than the load to be carried. And, light enough to be carried by the operator - making it fully portable. At this point, I revised the original morphological chart as follows:

Para-meters	Feasible sub-solutions				
Support	Wheels	Air cushion *	Tracks	Balloon	
Steering	Turning wheels	Air thrust	By hand *		
Stopping	Reverse power	Brakes	By hand *		
Moving	Air thrust	Power to wheels	By hand *		
Power	Electric mains *	Bottled gas	Electric battery		
Trans-mission	Hydraulic	Pneumatic	Belts	None *	
Lifting	Screw	Hydraulic ram	Pneumatic ram *	Belt	
Operator	Walking *	Remote control			

Morphological chart for a portable lifting vehicle (my own)

DISCUSSION OF ACTIVITY 20 (Continued)
Morphological Analysis (Forklift Truck)

As a result of a careful study of my own morphological chart for a 'portable lifting vehicle', I arrived at an unconventional vehicle configuration which I think is worthy of a design study (see table below). The sub-solutions I considered to be feasible and compatible (each marked with a * in the morphological chart) are as follows:

PORTABLE LIFTING VEHICLE	
Parameters	**Feasible sub-solutions** *My ideas:*
Support	air cushion
Steering	by hand
Stopping	by hand
Moving	by hand
Power	electrical mains supply
Transmission	none required
Lifting	pneumatic ram
Operator	walking

These functions, features and characteristics when combined into the total vehicle seem to me to add up to the following description of a new concept of lifting vehicle:

> ' A vehicle capable of lifting, hovering and transporting light loads. The air cushion and pneumatic lifting device being powered by an electrically powered air compressor. The vehicle being light enough to be carried by an average adult when not in use and easy to push and direct when moving loads.' Put more simply, it would resemble an electrically powered hover lawn-mower capable of lifting and transporting loads.

Having worked through this activity, you will have found that although there are 96,000 theoretically possible solutions to be arrived at from the chart, many of the combinations of sub-solutions are incompatible with each other. In fact, for design problems with more than a few parameters and sub-solutions, the number of unworkable solutions greatly exceeds the feasible solutions. However, I hope you found, as I did, that the use of a morphological chart can be a great stimulus to developing new design concepts.

IMPORTANT POINT The morphological chart should be used as a stimulus to developing new design concepts by exploring concepts other than those existing rigidly held combinations of sub-solutions.

WORK TASK 9

Generating alternatives by Morphological Analysis

This technique can be applied to either an existing product, or an entirely new product, process or service. The aim of Morphological Analysis is to generate and systematically review the widest possible range of alternative solutions.

If possible, this Work Task should be undertaken as a group activity. The group should try 'Brainstorming' the parameters, solutions and possible design concepts generated by the analysis.

1. Draw up a list of the basic parameters (i.e. *functions*, *features* and *characteristics*) you wish your product, process or service to possess.

2. Draw up a matrix, with the basic parameters down the left hand column, and use the boxes in each row to specify the possible different sub-solutions (i.e. the *means* by which each can be achieved).

3. Identify some *alternative design configurations* from the completed morphological chart. If the chart is small, you should be able to list all the possible alternative configurations. But if, as is more likely, the number of alternatives is quite large, then you will have to apply some selection criteria. Firstly, select some existing conventional *total solutions* from the chart - this should give you some practice at analysing the chart.

Having done this, select some unconventional and possibly novel design configurations by using the chart as a stimulus to new ideas.

The results of undertaking this task should be entered in your Work File for later discussion with your manager and / or tutor.

References

Pahl,G and Beitz,W, *Engineering Design*, The Design Council, London, 1984.
Pitts, G, *Techniques in Engineering Design*, Butterworth, London,1973.
Geschka, H, von Reibnitz, U and Storvik, K, *Idea Generation Methods - Creative Solutions to Business and Technical Problems*; Batelle Technical Inputs to Planning / Review No.5; Batelle Memorial Institute, Columbus, Ohio, 1982.

SECTION 3

CREATIVE APPROACHES
TO DESIGN PROBLEMS

STUDY GUIDE

In Section 2 of the Study Text you learned how to apply eight quite
different creative design techniques to the solution of design problems.
The usefulness of a particular technique depends upon the nature of the
problem to be solved and that stage of the design process to which it is
applied. In this final section of the book, I want to examine the problem
solving process in order to enable you to select the most appropriate
technique or combination of techniques.

The book begins by considering why creative techniques are needed and
how they can be used to stimulate ideas. We look once again at the *phases
of design convergence* described in Section 1 in order to judge where best
to use a particular technique and then look at the aims, advantages and
disadvantages of each technique. You are then asked to practise selecting
techniques for different design problems and processes.

OBJECTIVES
---

*After completing the study materials in this book you will be
able to:*

1. Explain the principles involved in using creative design
   techniques.

2. Understand how ideas can be stimulated by applying creative
   techniques to solving design problems.

3. Select a creative technique appropriate to a particular design
   problem.

WHY DO WE NEED CREATIVE TECHNIQUES ?

At the beginning of Section 2 we looked at the nature of creativity and some of the difficulties of deciding what it really is ! Whilst you may not agree with everything which has been said, we hope as a result of the activities you have undertaken, that you have been stimulated to think of ideas that you would not otherwise have had. The techniques you have used are all concerned with the decision making and problem solving processes of design. Design processes require *idea generation* by people who must by necessity be creative. We believe most people have the capacity to be creative and, given the right stimulus, most people can demonstrate creative abilities. Only you will know whether you consider your own creative abilities to be high or low. What we can say about the techniques presented is that they reflect the sort of creative thinking necessary to solve design problems.

The results of investigations into the thinking processes of highly creative individuals suggest, that when solving problems, they apply certain underlying principles. These include: abstraction, analogy, reduction, reversal, combination and variation. By using the creative techniques discussed in Section 2, we have shown how new ideas may result from the reversal, variation or new combination of existing solutions; through the transfer of ideas from previous experience to another sphere of activity; and through the use of analogies.

Creative people are seldom aware of the conscious application of these principles to their thinking processes. However, as a result of deliberately applying creative techniques to design problems, we know from the experience of those who have used them, and now hopefully from your own experience, that it is possible to stimulate new ideas.

HOW DO CREATIVE TECHNIQUES GENERATE NEW IDEAS ?

Creative techniques which generate ideas to solve design problems can be classified in several ways and according to two principles. The practical experience of a research group at Batelle Memorial Institute, Columbus, Ohio, USA, has found it useful to group the techniques according to their *working principles* and *triggering principles*.

Working Principles

The two working principles for generating ideas to solve problems are:

- Stimulation of the intuition

- Systematic approach to problem solving

Techniques which stimulate ideas in a spontaneous way involve the use of *intuitive thinking*. The Analogical Thinking technique is such a technique.

Techniques or methods which require an analysis and structuring of the design problem are *systematic approaches*. They involve a systematic analysis of the problem parameters in order to describe solution possibilities. The search for the size of parameters, often quantitatively, can lead to sub-solutions to the problem. Total solutions are then deduced through the combination and permutation of the sub-solutions. The Brainstorming and Morphological Analysis techniques are such approaches to problem solving.

Triggering Principles

The two triggering principles for generating ideas to solve problems are:

- Associations with, or variations on, other ideas or concepts

- Confrontations with oral or visual impressions

Associative triggering techniques encourage the development of chains of ideas through the involvement of members of an idea generation group. One such technique is Brainstorming.

Confrontation techniques generate ideas through the use of stimulating words or pictures and the subsequent forcing together of the ideas generated to solve the problem. The stimulation of new perceptions and impressions produce solution structures which can be transposed onto the problem to be solved.

These principles are relevant to the techniques you have used. The following chart shows how the principles relate to particular techniques:

Working Principle / Triggering Principle	Association / Variation	Confrontation
Stimulation of Intuition	Boundary Shifting Brainstorming Analogies	Boundary Shifting Brainstorming Analogies
Systematic Problem Solving	Morphological Analysis Objectives Tree Performance Specification	New Combinations Checklist

(Based upon the 'idea generation methods' classification developed by Batelle researchers. Reference: *Creative Solutions to Business and Technical Problems*; Batelle Technical Inputs to Planning / Review No.5; Batelle Memorial Institute, Columbus, Ohio, USA, 1982.)

WHERE DO CREATIVE TECHNIQUES FIT
IN THE DESIGN PROCESS ?

In Section 1 it was suggested that the design process moved through three phases, *concepts —> embodiment —> details* and, that two kinds of thinking are needed in solving design problems, *divergent* and *convergent*.

These ideas were also expressed in the form of a diagram, 'The Three Phases of Design Convergence'. This diagram is reproduced below, but this time indicating where in the problem solving and design process they are best used. Study this chart and read again the pieces on the **Design Process** and **Design Thinking** in Section 1 before going any further.

THREE PHASES OF DESIGN CONVERGENCE - SHOWING STAGES AT WHICH THE CREATIVE TECHNIQUES ARE USED

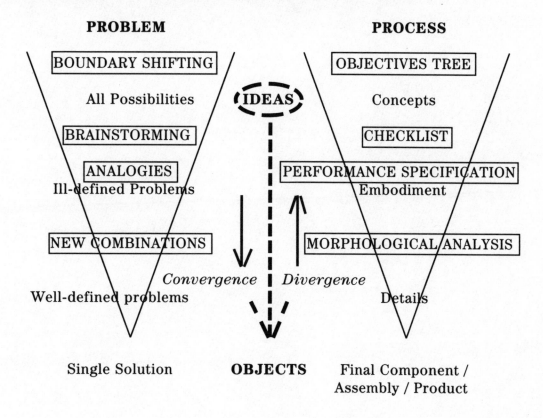

135

ACTIVITY 21
Selecting a Creative Technique for a Design Problem

In the table on the adjacent page you will see listed the eight techniques you have used, together with their main aims. By referring back to the Activities and Work Tasks in which you practised the use of each of the techniques, consider the advantages and disadvantages of each technique and enter your responses in your Work File.

Spend 30 minutes on this before comparing with my responses given in the Discussion of Activity 21 on the next page.

CREATIVE TECHNIQUE	AIMS
Boundary shifting	To provide new solution possibilities by shifting the constraints imposed by the problem
Brainstorming	To allow associative mutual stimulation within a group in order to generate many new ideas
New combinations	To look for new combinations of known forms and attributes
Analogies	To stimulate the thinking process through the use of creative associations and methods
Checklist	To stimulate ideas through the use of 'trigger' words
Objectives tree	To clarify customer requirements in design project situations at the start of the design process
Performance specification	To specify the most important requirements and features of the final product
Morphological analysis	To achieve new solutions by combining known sub-solutions in new ways

DISCUSSION OF ACTIVITY 21
Selecting a Creative Technique for a Design Problem

Below is a table which lists the eight techniques you have used and the aims, advantages and disadvantages of each. Your own thoughts about the advantages and disadvantages of each should be much the same, but you may have thought of others which apply in the context of your own design problem-solving situation. Deciding upon the usefulness of a particular technique or combination of techniques for solving your own design problems can only come about as a result of experience in their use. It is worth remembering that the skilful application of any technique is usually the result of practice - so do not quickly discard a technique without having tried it a few times.

CREATIVE TECHNIQUE	ADVANTAGES	DISADVANTAGES
Boundary shifting	Encourages a critical examination of the problem definition at the early stages of design. Moving problem boundaries can open-up or close-down solution possibilities.	Difficult to explain as an applied 'technique' to a particular problem - but rather as a discipline of exploration to follow.
Brainstorming	A quick way of generating ideas. Does not allow criticism. Spontaneous, uninhibited expression of ideas.	Criticism difficult to suppress. Experienced moderator essential. Group members tend to concentrate on own thoughts, rather than listening to others.
New combinations	New forms (or attributes) can be created by recombining existing ones. Encourages the analysis of design problems in terms of basic 'forms' and 'attributes'.	Difficult to decide in some cases what is a 'form' and what is an 'attribute'.
Analogies	Encourages an 'opening of the mind' to the possible contribution of disparate and accidental ideas.	Scope of ideas limited by the experience of the individual.
Checklist	Opens up new insights into design problems. Suggests different uses of existing ideas.	Ability to 'see' unusual combinations and oppositions.
Objectives tree	Reconciling the requirement of the market with those of the company's designers and manufacturing engineers.	Tends to encourage the use of the existing 'means' of achieving the project objectives.
Performance specification	Setting the parameters of the design problem in an unambiguous and precise way. Establishing the level of generality at which the design problem is to be solved.	Does not help in deciding how the design requirements are to be met.
Morphological analysis	Allows all possible solutions to a problem to be considered. Suitable for individual and group work.	Difficult to identify conceptual parameters. Complicated and time-consuming evaluation of large number of combinations of subsolutions.

WHAT ARE THE BENEFITS TO BE GAINED FROM USING CREATIVE TECHNIQUES TO SOLVE DESIGN PROBLEMS?

Most companies find themselves in situations of rapid change. These changes can be the result of the introduction of new manufacturing technology, new technology in the product or even the use of computerised technology in the process of design. Such changes are in turn driven by the demands of customers for having products updated, improved and delivered more quickly; and for more variety and better quality. All this is given even greater impetus by competitors trying to capture the market.

This situation requires that companies use a rapid and efficient product development process. The demands on designers are that they should be creative at all stages of the design process - in identifying customer needs, analysing the design problem and finding creative solutions. As a result of conventional education, most people have been schooled and are skilled in the use of logical thinking - what might be called *the straight line approach* to problem solving. This approach, whilst having its obvious merits, is in fact the antithesis of creative thinking. Traditionally, for instance in the manufacturing industry, a company's designers have been logical in their thinking, knowing that they are constrained by certain engineering principles, practices and laws. Problems have been solved by pursuing routine or known solution routes. But, in the present situation of rapid change, there is a need for new ideas.

Creative abilities need to be encouraged in all those actively engaged in the design process. People's potential to generate original ideas can only be discovered by stimulating their creative talents. This book provides an opportunity for creativity training by allowing individuals to explore their creative potential through fantasies, emotions and intuitions - without feeling the suppression of self-criticism. This is made possible through the application of individual or group creative techniques.

The benefits of such techniques can be listed as follows:

- Systematic but NOT logical problem solving
  - especially of ill defined and unstructured problems

- Use of time   —>   problem-solving period reduced
                       —>   better use through organised effort
                       —>   better applied thinking

- Ideas generated   —>   many more
                       —>   a more rational approach to selecting the best possible solution

- Groups —> can be used for all techniques if necessary
  —> benefits can be gained by including people from different disciplines.

- Individuals
  —> encouragement of communications skills
  —> self-expression and listening skills
  —> group integration and group dynamics
  —> improvement in risk taking by thinking about novel/unusual alternatives during decision-making
  —> encouragement to see colleagues as resources and not as rivals.

WORK TASK 10

Gaining benefits from creative techniques

Previous Work Tasks have given you practice in using the techniques you have studied. You have been asked to apply them to your own design problems and in the context of your own working situation. In this final Work Task, I would like you to carefully consider this experience and make suggestions for the continued application of suitable techniques to your own situation. This will mean considering two questions:

'What techniques are useful to me and how can I best apply them ?'

And, if you are employed by a company,

'What group techniques are useful to my company and how do I persuade my manager of their benefits ?'

(a) List the techniques you feel are useful, giving their aims and advantages, similar to the way you constructed the chart in Activity 21, but this time expressing your views in the context of your own design work.

(b) Using the list you made in (a), write a short report arguing the case for improving the process of generating ideas through the application of the techniques you recommend. You should be realistic by indicating the commitment and resources involved. At the same time, try to show the potential benefits to product design - through the use of the design work you have undertaken in previous Work Tasks.

Discuss this report with your tutor and / or manager.

A REVIEW

Having now reached the end of the Study Text, I suggest that you now spend some time reviewing its contents and the results of your activities. To help you with this, here is a complete list of the learning objectives that the study materials have been trying to help you achieve. Read through the list and ask yourself whether each objective has been achieved. If you feel that particular objectives have not been achieved, then I suggest you go back to the relevant section and study once again the parts in question.

CHECKLIST OF OBJECTIVES

You should now be able to:

Section 1 - The Nature of Design Thinking

1.1 Describe the general nature of the design process and outline its main phases.

1.2 Explain how different 'styles' of thinking are relevant in different phases of the design process

1.3 Identify your own preference for using a 'convergent' or 'divergent' thinking style

Section 2 - Creative Design Techniques

2.1 Describe some common features of the creative problem-solving process

2.2 Apply the following creative techniques.....
- Boundary shifting
- Brainstorming
- New combinations
- Analogies
- Checklist
- Objectives tree
- Performance specification
- Morphological analysis

Section 3 - Creative Approaches to Design Problems

3.1 Explain the principles involved in using creative techniques

3.2 Know how ideas are generated in applying creative techniques to the process of solving design problems

3.3 Select a creative technique appropriate to a particular design problem.

Suggested Readings

There are many books in design that take further some of the concepts and applications that have been introduced in these study materials.

A thorough and methodical design approach is explained very clearly in:

G. Pahl and W. Beitz, Engineering Design, The Design Council, London, 1984.

A more wide-ranging review of creative and methodical approaches to design is provided by:

J.C Jones, *Design Methods*, Wiley, Chichester, 1981.

A shorter book is:

G. Pitts, *Techniques in Engineering Design*, Butterworth, London, 1973.

A book which goes further into engineering science aspects of design is:

M.J French, *Conceptual Design for Engineers*, The Design Council, London, 1985.

If you are interested in product design, then a useful book is:

E. Tjalve, *A Short Course in Industrial Design*, Newnes-Butterworths, London, 1979.

Sources of Assistance

British Standards Institution
Linford Wood, Milton Keynes MK14 6LE
Telephone: 0908 320066

The Design Council
28 Haymarket, London SW1Y 4SU
Telephone: 071 839 8000

The Institution of Engineering Designers
Courtleigh, Westbury Leigh, Westbury, Wiltshire BA13 3TA
Telephone: 0373 822801